GEOCHEMICAL FACIES ANALYSIS

SERIES

Methods in Geochemistry and Geophysics

1. A. S. RITCHIE
CHROMATOGRAPHY IN GEOLOGY

2. R. BOWEN
PALEOTEMPERATURE ANALYSIS

3. D. S. PARASNIS
MINING GEOPHYSICS

4. I. ADLER
X-RAY EMISSION SPECTOGRAPHY IN GEOLOGY

5. THE LORD ENERGLYN AND L. BREALEY
ANALYTICAL GEOCHEMISTRY

6. A. J. EASTON
SILICATE ANALYSIS BY WETCHEMICAL METHODS

7. E. E. ANGINO AND G. K. BILLINGS
ATOMIC ABSORPTION SPECTROMETRY
IN GEOLOGY

8. A. VOLBORTH
ELEMENTAL ANALYSIS IN GEOCHEMISTRY
A. MAJOR ELEMENTS

9. P. K. BHATTACHARYA AND H. P. PATRA
DIRECT CURRENT GEOELECTRIC SOUNDING

10. J. A. S. ADAMS AND P. GASPARINI
GAMMA-RAY SPECTROMETRY OF ROCKS

Methods in Geochemistry and Geophysics

11

GEOCHEMICAL FACIES ANALYSIS

BY

WERNER ERNST

Geological and Palaeontological Institute,
University of Tübingen, Tübingen, Germany

ELSEVIER PUBLISHING COMPANY

AMSTERDAM / LONDON / NEW YORK

1970

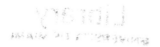

ELSEVIER PUBLISHING COMPANY
335 JAN VAN GALENSTRAAT
P. O. BOX 211, AMSTERDAM, THE NETHERLANDS

ELSEVIER PUBLISHING CO. LTD.
BARKING, ESSEX, ENGLAND

AMERICAN ELSEVIER PUBLISHING COMPANY, INC.
52 VANDERBILT AVENUE
NEW YORK, NEW YORK 10017

LIBRARY OF CONGRESS CARD NUMBER: 72-118252

ISBN 0-444-40847-9

WITH 34 ILLUSTRATIONS AND 18 TABLES

PRINTED IN THE NETHERLANDS

Contents

CHAPTER 1

Introduction

The geochemical analysis of facies is a special part of geochemistry and geology. The development of this research was initiated by BISCHOF (1847–1851), who was the first to attempt to recognize sediments of different origin by geochemical means. In his now famous book he listed some chemical elements belonging to typical environments, such as boron in marine sediments and metallic sulphides in zones rich in sulphuretted hydrogens. Moreover he was the first author to suppose that all "soluble rocks" of the continents must be present in the oceans. This early concept of facies was supplemented by ROTH (1879–1890) who discovered new elements in the ocean waters and in the ashes of marine organisms. However, the data used by both of these workers were quantitatively unsatisfactory and can only be considered as the qualitative results of inadequate chemical analyses.

Other analyses dating from the first decades of this century are also of questionable value for the diagnosis of facies. For example, CLARKE (1924) named only fifty elements in the last edition of *The Data of Geochemistry*; this is but half the number of elements known in rocks today. In the same work elements such as copper, lead, zinc and antimony could only be approximately determined. These and other trace elements are, however, very important for geochemical facies analysis in anaerobic environments. The progress of research was therefore dependent on the development of chemical techniques, and also on the concept of facies drawn up by geologists after the special facies session of the Geological Society of America in November 1948.

For these reasons modern geochemical analysis was only possible after the last war. Since that time a great number of geochemical studies of facies have been published, but they are dispersed in many small papers all over the world. Most of these published results are collected here in this special treatise on geochemical facies analysis. The book itself is based on the concept of facies analysis formulated by KREJCI-GRAF (1966).

CHAPTER 2

The Meaning of Facies

Each branch of the natural sciences has a special expression for the influence of physical and chemical forces on a defined body. In the earth sciences this term is "facies", and since its first application by Steno in 1969 it has undergone many changes. In the last century Gressly in 1838 used this concept in differentiating fresh, brackish, and marine sediments in the Jurassic beds of eastern France. He regarded all geological, petrographical and palaeontological features of the rock as facies criteria. At about the same time Prevost distinguished pelagic, subpelagic and littoral facies zones, as well as coralline and sponge facies. His facies concept thus embraced the sediments as deposits as well as their chief organic components. Suess applied the concept in the same sense.

All the facies definitions so far cited related to the horizontal dimension. GOLOVKINSKII (1869) introduced the vertical component into the concept meaning thereby "the migration of facies in time and space". WALTHER (1893), in contrast, designated only the characteristics of contemporaneously formed sediments as facies. As criteria Walther took into consideration solely the lithological characteristics of the sedimentary province. His most important contribution to the knowledge of facies is the "correlation of facies".

Of the many facies definitions coined during this century few are suitable for geochemical purposes. The concept of KAZAKOV (1939) is quite without relation to the conditions of deposition for he understood under facies only the distinction of palaeogeographical and geochemical characters. The same author also separated "recent" and "ancient" facies, and designated as recent facies a continuous region of the earth's surface distinguished through the uniformity of physio-chemical or geochemical features. PUSTOVALOV

(1940) took over the essential parts of this definition for the distinction of "ancient geochemical facies". The facies thereby relate to single beds, or a sequence of beds, which during their formation were exposed to homogenous geochemical conditions over considerable distances (e.g., marine facies). Pustovalov divided facies into further sub-units distinguished within a particular facies unit through certain peculiarities. Hydrosulphide and siderite facies characterize, for instance, different oxidation potentials within the marine facies.

As a further unit Pustovalov distinguished the "terrigenous-mineralogical facies" which took into account not only the differing provenance of the minerals but also the change in the terrigenous components during the process of sedimentation. The attempt was thereby made to bring the origin of the material into the facies definition. VASSOYEVICH's (1948) consideration of the changes of the facies characters in the separate developmental stages after deposition appear to be equally important for geochemical purposes. Vassoyevich distinguished:

(1) Origofacies: the facies of the primary depositional environment.

(2) Lapidofacies: the facies of the diagenetic environment.

(3) Densofacies: the facies of the metamorphic environment.

(4) Exedofacies: the facies of the weathering environment.

These facies definitions incorporate the sum of all the rock characters both during and after lithification. The concept of facies itself distinguishes in particular the supplementary changes which could obliterate the original aspect of the facies. A further differentiation of facies was suggested by RUKHIN (1948). It was based on the distinction of lithologically identical deposits characterized by the same conditions of deposition (e.g., sandy facies). Several such genetically similar facies units were summarized under the name macrofacies.

At the same time KLENOVA (1948) redefined the term marine facies in a new sense: in his opinion the marine facies could be distinguished through the constancy of physico-geographical (or physico-chemical) and biochemical relationships as well as through a common source of sediment.

Most of the definitions discussed up to now have been related only to lithological features and special circumstances of sedimentation. MOORE (1948, 1949, 1957) introduced a further facies definition which is exclusively applicable to the development of the inorganic components of sedimentary beds. This complex he labelled physiofacies. Moore distinguished in addition between a facies which represented a separable part of a stratigraphical unit, and the lithofacies. The lithofacies is a unit quite independent of the stratigraphy, embraces all the rock characters (inorganic and organic) of a sedimentary deposit, and indicates the circumstances of its deposition. The lithotope, introduced by WELLS (1944) for the sedimentary development (record) of a biotope, had in Moore's view relevance only to the actual environment and not to its manifestation as rock.

Correspondingly the term biofacies denotes the total biological characteristics of a sedimentary deposit. SLOSS et al. (1949) used this term for faunal or floral units whereas KRUMBEIN and SLOSS (1951) used it to signify the fossil development of an organic environment (biotope). Another term was established by SLOSS et al. (1949, p.96) for "a group of strata of different tectonic aspect from laterally equivalent strata". This was designated tektofacies. Correspondingly the same authors recognized a tektotope as "a stratum or succession of strata with characteristics indicating accumulation in a common tectonic environment".

All the definitions which concern litho-, bio-, and tektofacies have undergone certain modification at the hands of American geologists. DUNBAR and RODGERS (1957) put all lithological and, in a broad sense, all structural, tectonic and even metamorphic characters in their facies concept. LOCHMAN (1956) used the expression facies for undefined rocks which could only be distinguished on the basis of their lithological features. KUMMEL (1957) interpreted facies as tongue shaped "bodies" which are lithologically determinable. These facies definitions, resting only on lithological characters have no essential importance for the diagnosis of geochemical facies. The same applies to the commonly quoted terms: geosynclinal facies, platform facies, fore-deep facies, which were erected by SLOSS et al. (1949).

Thus of the numerous definitions only a few can be considered for geochemical purposes. The geochemical facies of PUSTOVALOV (1940) requires constant geochemical conditions over considerable distances, rarely realisable for other than marine deposits. The further subordinate facies units of Pustovalov (e.g., hydrosulphide and siderite facies) are applicable only to marine sediments, although different oxidation potentials may occur in fresh water. Pustovalov's terrigenous-mineralogical facies include two totally different aspects of facies, namely both the provenance and the change of the sediment in the depositional province. Moreover, this term is easy to confuse with the common expression "terrestrial facies", which relates to continental deposits, e.g., coal formations.

Moore's physiofacies is too broadly based for geochemical purposes, since in this concept all inorganic components of a deposit are included: minerals, trace elements bound on to clay minerals, and pore water. Consequently physiofacies can only be considered as a kind of super-unit.

Therefore as a special facies unit for geochemical purposes, the hydrofacies is here proposed. Under the geochemical characters salinity, temperature and redox-potential are understood; all of which are influences capable of creating a special facies picture in the sediment, and which are not exclusive to a depositional domain such as the sea or the land. As subordinate units salinity, temperature, and oxygen facies, of which the first is already in the literature, come into consideration.

Geochemical facies analyses, which are connected with rock properties and the organic content of the rock, are consequently combined with the lithofacies. Of this kind are facies analyses of limestones, dolomites, and reef complexes. Facies properties based totally or dominantly on organic components are assigned to biofacies. Among these are diatomite, radiolarite, flint, and coal deposits.

These expressions are also retained even if the investigated rock has suffered metamorphism or weathering, so long as characteristics of the facies have not been erased.

CHAPTER 3

Prerequisites of Geochemical Facies Analysis

PERMANENCE OF THE OCEANS

One of the most important prerequisites for the repeatability of geochemical facies analysis is the permanence of the composition of the oceans through long periods of the earth's history. This permanence can only be maintained if the introduction through terrestrial weathering and exhalation of the chemical components dissolved in the water balances their removal by organisms and sedimentation. It is also conceivable if increasing erosion or exhalation can be compensated by a continuous growth of the water body. The balance between introduction and fixation can, however, be considerably disturbed if, for example, large quantities of salt are removed from the ocean in a relatively short period of time.

Less essential preconditions for sediment formation are water and weathering processes. Weathering can occur under reducing (Fe, Mn) as well as oxidising conditions, and is therefore independent of the original composition of the atmosphere; in both cases, however, water is necessary as an agent of solution and transport. Somewhat diverse opinions exist over the stage in the early history of our planet when the condensation of water took place. RUBEY (1951, 1955) maintained it was possible without the necessity for any essential change in a primeval atmosphere of carbon dioxide, nitrogen and hydrogen sulphide. On the other hand UREY (1951a, 1959) believed that the early atmosphere was composed of methane, ammonia, hydrogen, water and nitrogen (plus subordinate carbon dioxide and hydrogen sulphide). In the case of this model the hydrosphere was probably able to develop only after the photochemical dissociation of the water in the higher atmosphere. Prerequisite were also the gravitational separation of hydrogen and oxygen in the outer zones of the atmosphere, and the partial oxidation of CH_4, NH_3, H_2S, C, N_2, and Fe^{2+} (WEDEPOHL, 1963). In this interpretation the primitive atmosphere must already have been considerably altered before the formation of the water masses. Accordingly the condensation of water must have taken place later than in the case of Rubey's model. In favour of the Urey/Wedepohl model is, on the one hand, the very considerably reduced hydrogen content of the atmosphere at the present day, which accounts only for a tenth of the oxygen content of the atmosphere, in contrast to the primeval state. On the other hand, the origin of life demands just those substances, such as methane and ammonia (OPARIN, 1957), which are assumed in Urey's model atmosphere.

The original quantity of water in the primeval ocean is difficult to estimate. Owing to the high ground temperatures then current a considerable part must have existed as steam; the remainder was probably distributed in several basins. In comparison to the other components of the primeval atmosphere water has a markedly earlier condensation point (boiling point of NH_3 and H_2S: 33.4 and 60.75°C respectively). It could therefore have been stable in a stage of the geochemical development process in which the overall physio-chemical reactivity was several times greater than at essentially lower temperatures.

Additional favourable factors are the polar properties of the water molecule which considerably ease the building of aggregate molecules, and its high dielectric constant. These properties of water and the relatively high temperatures are already sufficient to cause a division of many materials into ions. This may have resulted in an intensification of the weathering which had probably begun long before the primitive components of the atmosphere and their oxidation products (dominantly CO_2) were dissolved in quantity in the ocean. This weathering was considerably strengthened by the formation of the aggressive $(HCO_3)^-$ ion in the water. Through the intense silicate weathering thus introduced the supply of alkalis and earth alkalis must have been considerably increased.

Similarly under the reducing conditions of the inceptive stages of the primeval atmosphere and the primitive hydrosphere, bivalent iron as well as manganese must have passed preferentially into solution, and this in turn must have led to a reaction with the H_2S of the sea water.

Under these restrictions the early sediments of the primeval ocean would have been Mg-rich limestones and Mn-rich iron sediments which are scarcely likely to be preserved today. With the introduction of weathering a part of the gaseous components of the primeval atmosphere was lost through reaction with the substances thus introduced. A further part was used in the formation of the first organic substances. This latter process may have followed relatively early in the development of the primeval ocean. In the opinion of OPARIN (1957) before the actual development of primitive life, organic combinations of carbohydrates must already have been in existance. They were in part present in the primeval atmosphere (Urey) and were dissolved in water as the non-oxidized remnants of this primeval atmosphere. Condensation of these carbohydrate molecules is easier to imagine in narrowly restricted bodies of water.

Other possibilities for the creation of the basic material of living organisms were already set in motion in the primeval atmosphere: the decomposition of NH_3, CH_4, H_2O through (UV) radiation and the formation of formaldehydes. Many experiments exist concerning how the decomposition products can react with one another, and how formaldehyde in aqueous solution condenses in the presence of calcium (glycoaldehyde, fructoses, hexoses). In this manner the sugar complexes, amino acids, purine and pyrimidine bases, which under the thermodynamic conditions of the primitive earth surface were capable of forming polymers, can be explained.

In summary: weathering was possible immediately after the first condensation of the primeval atmosphere and was considerably enhanced by the gradual incorporation and mixing of the constituents of the atmosphere in water. The gases of the primeval atmosphere when dissolved in water were used up both through reaction with weathering solutions, and through the development of organic substances. The condensation of this originally sparsely distributed organic material was favoured on the one hand by a limited volume of water, and on the other by the presence of concentrated calcium solutions. We can therefore assume that this first biochemical phase must have taken place in a relatively early stage of the primeval ocean. This phase practically closed with the consumption of most of the primary components of the condensed atmosphere, and the end of the intensive silicate weathering.

Through the work of CONWAY (1942, 1943), KULP (1951), TUREKIAN (1959), BRAITSCH (1963), and WEDEPOHL (1963), we are considerably better informed about the second phase of the development of the oceans. In the first phase of ocean development the reaction of the water with the substances dissolved in it had primary importance. In the second phase the degasification of the earth is, indeed, the essential source for the further increase of sea water in early times. It yielded, in addition, the greater part of the CO_2, Cl, S, N_2, and B in the course of this stage of development. The most volatile constituents (Rubey: excess volatiles) cannot be accounted for solely by weathering according to the balance investigations made by GOLDSCHMIDT (1933), CONWAY (1942, 1943), CORRENS (1948), and WEDEPOHL (1963) one must conclude that degasification in the course of the later history of the earth decreased considerably and has remained more or less the same for a considerable time up to the present day.

The last, and for geochemical facies diagnosis, most important

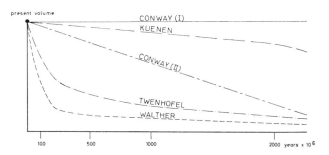

Fig.1. Increase in ocean water (Mason) and balances from Conway, Kuenen, Twenhofel and Walther. (After MASON, 1962; with kind permission of the publisher.)

phase in the development of the oceans began with the influx of cyclic sedimentary material, and the proper onset of organic life. In this period from the close of the Precambrian to the present day the water in the oceans and sediments has increased according to the estimates of Turekian by only about 15% (through degasification, weathering of sediments, metamorphosis). The increase in water volume in the view of this and other authors is represented in Fig. 1. Weathering resulted in the arrival of dominantly reworked sedimentary material in the basins. In the previous phases, in contrast, eruptive rocks were involved. With the transfer of weathering from eruptive to sedimentary rocks the quantity of dissolved material transported in rivers probably increased threefold (WEDEPOHL, 1963). Thereby more Ca and considerably less Na reached the oceans.

The biogenic influences on the composition of the sea water are also very important. Through assimilation CO_2 is used and O_2 set free. In sea water O_2 is lost both through breathing and through the oxidation of dead and sinking plankton. How much O_2 is consumed in this way can be estimated from the annual production of 10^{12}–10^{13} tons of planktonic organisms (Wedepohl following calculations by SKOPINTSEV and TIMOFEYEVA, 1960). Similarly CO_2 is extracted for the formation of calcareous skeletons. Animals with calcareous shells also absorb in this way Sr, B, Cu, Mn, Fe (PROKOFIEV, 1964; KREJCI-GRAF, 1966). Aragnote shells may concentrate B, Ba, Rb, and Sr, depending on the facies conditions (LEUTWEIN, 1963). In comparison with sea water marine organisms are enriched to the order of 100–4,000 times in Cu, Fe, V, Zn, and some in Au (KREJCI-GRAF, 1966). Notable quantities of H_2S arise during the decomposition of organic substances (proteins) under the influence of bacteria in sediments. H_2S reacts to some extent with heavy metals and builds protective zones against further oxidative decomposition of the dead organisms in isolated regions of the sea. Nitrogen originates both from the decomposition of protein (protein→amine→NH_3) and from the degasification of the earth's mantle. Similarly the consumption of CO_2 through assimilation and the formation of sediments and calcareous shells is compensated by the supply from the earth's mantle. Analogous postulations are applicable in RICKE's (1963) opinion to sulphur.

Further components of the ocean water are bound on to clay minerals. Newly formed chlorite (JOHNS, 1963) contains about 4% water. In a like manner montmorillonite takes up water between its layers. Considerable quantities of water can be bound up in unlithified sea floor mud. More important for chemical facies analysis, however, is the adsorption or total incorporation of trace elements in clay minerals. In the formation of chlorite from montmorillonite chlorine is built in (JOHNS, 1963). Boron can probably substitute for aluminium in the tetrahedral position in micas (HARDER, 1961). A not inconsiderable number of trace elements can also be bound on to the organic substances of bituminous sediments.

On the other hand trace elements are set free in solution through the reworking of clay sediments during transgressions. Similarly trace elements get into both the surrounding sediment and sea water from decayed hard parts of organisms. A few decades ago it was still supposed that the progressive loss of trace elements in sediments must generally impoverish the sea water in these elements. GOLDSCHMIDT (1954), for instance, thought he recognized such a trend in the boron content of the sediments he compared. On the contrary trace elements may be concentrated in evaporating regions of the sea, so that an enrichment in the sediment must occur with decreasing geological age (e.g., the distribution of bromine in sylvine). An important question is, therefore, how these processes, restricted spatially and to separated periods of the earth's history, affect the overall distribution of the elements.

There is still no adequate experimental material which comprehensively records practically all changes. A very rough view of the distribution of elements in marine clays of varying age in the Northern Hemisphere is shown in Fig.2. There are differences there between the figures for boron given by GOLDSCHMIDT (1954) and ERNST (1962, 1966), which are traceable to the errors present in early analyses (ERNST et al., 1958). The recent investigations of HARDER (1963), PORRENGA (1963), and ERNST (1966) speak for a relative constancy of the content of boron over long periods of

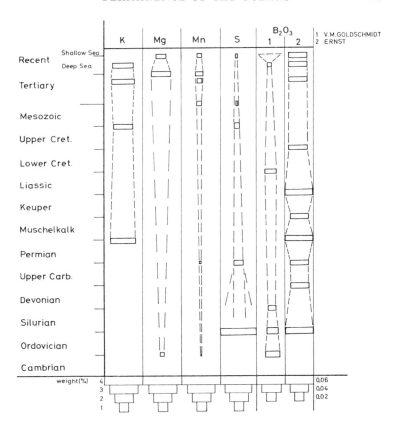

Fig.2. Average value of selected elements in various geological systems. (After GOLDSCHMIDT, 1954; ERNST, 1962, 1966; LANDERGREN and MANHEIM, 1963; RICKE, 1963.)

time. The figures of RICKE (1963), and LANDERGREN and MANHEIM (1963) show a tendency for the sulphur content to decrease after the Ordovician. In contrast Mn and Mg increase to the present day, whilst Fe (total iron) remains approximately the same. This balance can therefore only illustrate tendencies for individual elements. On the other hand a survey of the various chemical elements on the Russian platform from the Proterozoic to the Quaternary is based on considerably more analyses (VINOGRADOV and RONOV, 1956).

TABLE I

DISTRIBUTION OF EVAPORITE DEPOSITS IN TIME AND SPACE
(After BRAITSCH, 1963)

Deposit	Location
Upper Tertiary	southwestern Siberia, Iran, Galicia (foreland of Carpathians), Azerbaijan
Lower Tertiary	central Asia, upper Rhine (France, Germany), Ebro (Spain), Somali (Africa)
Cretaceous	Louisiana, Texas, Florida (U.S.A.), Tadzhikistan (U.S.S.R.), South America, central Africa
Upper Jurassic	Arkansas, Alabama, Utah, Texas (U.S.A.), Tadzhikistan, northwestern Germany, western Europe
Triassic	central and western Europe, Sahara (North Africa), Apennines (Italy)
Upper Permian	central Texas, Colorado, New Mexico, Utah (U.S.A.), northern Germany, Great Britain, Poland
Lower Permian	Louisiana, western Kansas, Oklahoma (U.S.A.), western foreland of the Urals, Caspian Depression (U.S.S.R.), Austria
Upper Carboniferous	Colorado, Utah (U.S.A.)
Lower Carboniferous	Dakota, Michigan, western Virginia (U.S.A.), Canada, Amazone district
Upper Devonian	Dnieper-Donetz Basin (U.S.S.R.)
Middle Devonian	Alberta, Michigan, northern Dakota (U.S.A.), Saskatchewan (Canada)
Lower Devonian	northern Siberia (U.S.S.R.)
Silurian	Michigan, New York, western Virginia (U.S.A.), central Siberia (U.S.S.R.)
Lower Cambrian	Angara-Lena (U.S.S.R.), Punjab saltrange (India), Hormiz series (Iran)

This survey showed K to decrease through the whole succession and Al and Ti in the Mesozoic, whereas the Ca/Mg ratio increases from ancient to young sediments. The maxima of the Al and Ti contents are concurrent with times of transgression or the formation of continental, littoral or coal beds. For Sr a review of North American limestones is available, according to which the Sr content decreases from the Ordovician to the Pennsylvanian (KULP et al., 1952). Ca shows a tendency to increase as the earth's age decreases (WEDEPOHL, 1963).

These reviews reveal that no constant relationship exists for individual elements in the last phase of ocean development. This applies to all the subordinate constituents of sea water which come in fifth place or later after H_2O, Cl^-, Na^+, and $SO_4{}^{2-}$. The deviations in their concentration can be explained through fluctuating weathering intensity and diverse transgressions. According to the above surveys the considerably smaller variation of the so-called volatile elements is due to a compensatory balance between supply (degasification) and incorporation in the sediment. On the other hand the chief components of the ocean which occupy the first five positions in the concentration scale suffered much greater changes in their concentrations since they were involved several times in the formation of salt bodies. Table I shows the most important salt deposits of the earth more than 100 m thick. The quantity of salt held in these concentrations is not, however, exactly known. From an estimate made by RICHTER-BERNBURG (1955) modified by BRAITSCH (1963) $2 \cdot 10^{14}$ tons NaCl ($= 0.5\%$ of the NaCl content of the seas) can be allotted to post-Variscan Europe alone. A rough estimate by BRAITSCH (1963) assigns a total value of $2 \cdot 10^{15}$ tons for all the salt deposits of the world ($=$ approximately 5% of the NaCl content of the seas). The Cl-content of this salt amounts, according to BRAITSCH (1963), to $1.2 \cdot 10^{15}$ tons in comparison with $0.13 \cdot 10^{15}$ tons in the remaining sediments (excluding deep sea zones). In contrast the figures for $CaSO_4$ are approximately $7 \cdot 10^{15}$ tons for sediments (AULT and KULP, 1959) and $7 \cdot 10^{15}$ tons for evaporites (calculated from average given by RICKE, 1963).

Thus the differences between the quantities of sediments and evaporite bound NaCl is small in comparison with the reserves in the oceans. The quantities of $CaSO_4$ are, on the contrary, of about the same order of magnitude as the available reserves in the sea. The Ca^{2+} and $SO_4{}^{2-}$ content may therefore have varied and fallen after each period of salt formation without the chlorinity and alkalinity of the sea water having been thereby altered.

Thus in conclusion one can say that during the third phase of the development of the ocean water volume, supply and elimination of chemical components remained, on average, stable. Above all the subordinate non-volatile constituents underwent fluctuations, and to some degree also Ca^{2+} and SO_4^{2-}. For geochemical analysis, however, these elements are of limited significance. Much more important are most of the volatile components which in this phase of ocean development remained in equilibrium between supply and fixation. The variations in concentration in the two earliest phases have no essential significance for geochemical facies analysis since practically no sediments of this age have been preserved.

CLIMATIC INFLUENCES

In comparison to the long term changes in the composition of the oceans the climate has undergone frequent variation. Weathering and the formation of special sediments on the continents and in marginal basins of the sea are most strongly affected by such variation. Particularly active climatic periods in the earth's history are relatively easy to read from the evidence of their special products (evaporites, coal beds, tillites).

With certain exceptions evaporites originated in times of relatively dry and warm climate. They hold, as we have seen, only about 5% of the NaCl of the world's oceans, and $CaSO_4$ in about the same order of magnitude as its marine reserves. The accessory constituents of the oceans, however, are considerably more affected by this evaporation. Mg is incorporated in dolomite, magnetite, hydromagnetite, Mg-hydrogen-carbonate, and newly formed chlorite, whilst on further evaporation Ca reacts with the sulphates of the sea water producing gypsum (BRAITSCH, 1963). Boron is enriched in the anhydrite of the centre of the German Zechstein basin to a level of 400–600 g B/ton as compared with about 20 g B/ton on the margins (after Harder in BRAITSCH, 1963). Strontium (diadoch for Ca in anhydrite; celestine) also increases with the precipitation of rock salt but achieves its maximum on the continental borders. Bromine becomes enriched in the final liquor of evaporite basins and consequently achieves its highest concentration in the youngest beds of potassic precipitate sequences (diadoch representative for Cl in sylvine).

In the event of an inflow of sea water and an outflow of liquor into the sea the precipitated chemical combinations are quickly replaced. On the other hand in the case of salt deposits formed from sea water cut off from the ocean these elements become impoverished. We must, for instance, reckon with a smaller content of boron in the younger beds of such a basin that at the commencement of salt precipitation. In other salt basins, characterized by the mixing of remnant oceanic water and continental solutions, an excess of Na_2SO_4 is dominant.

A hot, dry climate leads to special weathering phenomena on the continents. In rain-free or -poor areas of this climatic zone the mineral framework decays and weathering residues are produced without being carried away. During occasional soaking of the ground evaporation causes the formation of efflorescences on the surface (salt blooms). During an alternation of dry and wet periods in a warm climate periodical precipitation of the chemicals dissolved in the ground water or within reach of the surface occurs (calcareous, siliceous, gypsum, salt and nitrite crust formation). Red earth is a product of subtropical and tropical zones with a warm and periodically wet climate. SCHWARZBACH (1961) regarded them as typical for present day savannahs, but fossil examples must have extended over poorly vegetated areas affected by alternations of humidity and aridity (Rotliegendes, Late Permian, Triassic). The red colour of this sediment is often secondary. In the case of the central European Rotliegendes it has been derived from weathered Carboniferous beds.

Cu is a typical enrichment product of such red sediments. Six copper-bearing horizons connected with red sandstone lenses containing plant remains in the Permo-Trias of Marocco have a Cu content of about 0.7 %. Other Cu concentrations in arid areas are known from the foreland of the Urals and in the United States. The Rotliegendes of central Germany yields Cu only in the order of 10^{-3}. This quantity is insufficient for the derivation of the metal content of the German Kupferschiefer. Most sediments are very rich in boron. The B content of red clay sediments is two or three times in excess of that in marine shales (ERNST, 1962). Since boron is very stable during weathering (because of the strong link to the mica lattice) its concentration is likely to have been produced by the evaporation of a shallow sea and not by its having been carried in ascending solutions. These boron-rich sediments commonly include anhydrite particles between the clay minerals, which indicates diagenetic change after the laying down of the fine clay. Plants have not yet been found in the neighbourhood of anhydritic clay stone, but they do occur in the Stephanian and Lower Rotliegendes where they even formed coal seams syngenetically with the red sediment. In contrast the red colouration in grey coal-bearing successions, e.g., in the Mid Coal Measures of south Wales (BLUNDELL and MOORE, 1960) and in the Carboniferous of the Ruhr (SEIDEL, 1962) can be explained as post-sedimentary. Coals are not, however, associated only with red sediments, but occur together with indicators of a cooler climate: in Australia the lowest coal seams are found immediately above a series of tillites; in South Africa boring has revealed thin coal seams clearly intercalated in the uppermost zone of a tillite more than 300 m thick (PLUMSTEAD, 1961).

The same author recorded the presence of *Gangamopteris* and *Glossopteris* both before and during the glaciation. The tillites, as well as deciduous plants and trunks with annual rings, hint at a cool climate during the formation of the Late Carboniferous coals of the southern continents. The red sediments with autochthonous coals indicate, in contrast, at least a warm if not a hot climate; in both cases rain was evidently periodic. These examples lead to the conclusion that coal beds are possible in a wide range of climatic zones and are not exclusive to warm, humid areas (Late Carboniferous coal belt). Their value as an indicator of climate is therefore restricted. Similar restrictions can be raised for the evaporites of small depositional troughs. For example, the clay layers of the Oligocene potassic deposits of the upper Rhine Graben yield well preserved plant remains which were indigenous to tropical forests not far from the salt-basin, and indicate a wet climate.

Fossil laterites (bauxite), marine limestones and reefs are, however, certain evidence of a warm climate. Kaolinite and kaolinite seat-earths of brown coal seams speak, under certain conditions (a granite hinterland), for a warm, wet climate. Cold climates in the geological past are incontrovertably demonstrated by such features as erratics, abraded surfaces, moraines and glacial landscape forms. The above examples have shown that not all so-called climatic indices permit a certain diagnosis of the climate. If we wish to judge the influence of the climate on the composition of the ocean water then we must restrict ourselves to the more or less contemporaneous climatic phenomena which are distributed over very wide areas of the world.

In the case of salt we know of deposits from the Silurian, Middle Devonian, the Permo-Triassic, the Late Jurassic and the Tertiary (see Table I). We know from the calculations of BRAITSCH (1963) that these precipitates had no essential influence on the chemical composition of the world's seas. On the other hand the peripheral seas were much more strongly affected by changes in chemistry, becoming impoverished in minor constituents if no substitution took place through an influx of sea water. Geochemical facies analysis in such situations must be made with this in mind.

Peripheral basins cut off from the ocean also serve as temporary traps for components introduced by rivers. The chemistry of their water often differs from that of the open sea and the salts deposited in them are lost to the ocean water for a considerable time. Not only does this result in sediments differing from those of a fully marine regime but special faunal communities also arise (brackish water fauna of the Caspian Sea). Such basins on the continental margin in a dry, hot climate are imaginable during several periods in the history of the earth: in the time of the Old Red Sandstone, the Rotliegendes, Buntsandstein and Keuper.

Periods of coal formation often precede those during which salt was deposited:

Coal	Salt
Late Carboniferous	Permian
Late Keuper	Middle Keuper
(Lettenkohle in Germany)	(Gipskeuper in Germany)
Tertiary	Tertiary
Eocene	Oligocene
Miocene	Miocene/Pliocene (Carpathians)

One can visualize a connection between the elimination of CO_2 during the early stages of carbonization and therefore an increase in the CO_2 content of the atmosphere (green-house effect). Investigations carried out by CLOUD (1952) and PLASS (1956) (quoted in WEDEPOHL, 1963) support ARRHENIUS's (1950) opinion that the climate appears to be strongly dependent on the CO_2 content of the atmosphere, an increase being concomitant with a rise in the temperature of the atmosphere. As earth gas measurements have recently shown (ERNST, 1969), gases can escape in notable quantities from beds near to the surface through the opening and closing of expansion joints under the influence of the daily rythm of the moon (earth tides or the gravity influence of the moon). Through the very widespread distribution of coal-bearing beds in the Late Carboniferous, the carbonization of which began in Westphalian D and at the latest before the onset of the Zechstein transgression, relatively large quantities of CO_2 could have been released into the atmosphere. Similar conditions result from the formation of oil concentrations near the surface of the earth.

The coal basins could have acted, like the salt basins peripheral to the oceans, as traps for chemical components introduced by river water. In the Late Carboniferous of Westphalia it is known that the rivers in the region of the coastal marsh often meandered strongly. They branched into many canals and often flooded over their own banks to form large seas over the vegetated area. The plant ash of the coal seams consequently contains numerous trace elements typical for the coal, and often considerable quantities (Ge, Be, As, Cu, Zn, Pb).

These elements are also lost to the peripheral seas (fore-deep seas) until the sediment is reworked by a new transgression. In the total balance of the oceans, however, such losses are hardly noticeable. During dry periods, and the times of coal formation which may occur at various temperatures, a local restriction of transport must therefore be reckoned with. The sediment only physically weathered during a hot, dry climatic interval and temporarily resolidified as crusts, can be introduced into the marine domain by a transgression. Thereby large quantities of accumulated chemical components are suddenly released in the same manner as during the reworking of a thoroughly weathered rock (bauxite formation). VINOGRADOV and RONOV (1956) showed that the maxima of Al and Ti are to be found in the beds resulting from such transgressions. In this case the climate is of considerable importance in geochemical facies analysis.

Summing up one can say that the short-lived, periodic influence of the climate made little impression on the composition of the ocean waters but had somewhat more effect on the waters of peripheral basins. Coal and small evaporite basins are uncertain climatic indices for their regional environment, since they can be formed in a variety of climatic zones. They should therefore not be used as standards in judging geochemical climatic data.

TECTONIC INFLUENCES

The intensity of weathering and the transport of solutions in rivers is dependent not only on climate but also on tectonic influences. The formation of peripheral seas (Black Sea, Red Sea) and the growth of fore-deeps in front of rising mountains are also governed by tectonics, as are the marine transgressions, which penetrate far on to the land in platform and shelf areas.

The intensity of weathering is a question of climate and the site of the weathering surface. Differences in altitude of several hundred metres may occur between water level and the surrounding mountains in a narrowly restricted region, and a uniform climate can result in totally different weathering intensities. In inter-montane Carboniferous basins (internal molasse) a warm periodically wet climate led to profound red weathering in the mountains, whilst contemporaneous coal formation was possible in the deepest parts of the basin. Under identical climatic conditions tectonic influence is thus able to produce very considerable differences in the character of the weathering.

Which rocks appear in the weathering zone is also subject to tectonism. In the case of the Bohemian Massif Granite was brought into the zone of erosion by the Variscan orogeny, and a warm, wet climate on the Triassic–Jurassic boundary caused intensive weathering and kaolinisation of the granite. Consequently we find a remarkable number of kaolinite minerals in the terrestrial deposits of the Jurassic on the western margin of the massif (KRUMM, 1963). The large kaolin content disappears in favour of illites only in the zone transitional to marine Jurassic beds.

The type and intensity of the tectonic movements govern the removal of the weathered material. In flat-lying regions material weathered from granite masses cannot be transported away (Finland, southern Sweden), but in mountain areas blocky detritus can be moved to accumulate in broad fans on the sides of the valleys. Thus during the early stages in the development of a fore-deep only coarse material reaches the depositional trough, and the sedimentation thickness is correspondingly very great. With retreat of the erosion rim or extension of the sedimentary basin clay material can also be washed into the zone of deposition accompanied by a lowering of the rate of sedimentation (mature fore-deep, e.g., sub-Variscan fore-deep).

These apparently trivial relationships are in fact of considerable importance for geochemical facies analysis. One cannot hope to find the trace element content associated primarily or secondarily with illite in a sediment which is composed dominantly of kaolin. In rapidly accumulated sediments there is relatively little clay or organic substance, and moreover in such depositional zones the incorporation of trace elements in the clay minerals is constantly inhibited by disturbance of the sediment (this will be discussed in more detail in the following section).

The creation and development of fore-deeps and peripheral seas is a further area in which tectonism plays a vital role. The fore-deeps created in the course of an episode of folding receive practically only the sediment found in the immediate hinterland. Each fore-deep therefore shows a somewhat different sedimentary assemblage, which impedes geochemical facies analysis. The division of peripheral seas by swells and basins is largely responsible for salt formation under dry, warm conditions, and also for the oxygen-poor (euxinic) milieu in deep water in all climates. In this way special facies conditions are created which can only be dealt with by special geochemical methods (oxidation facies).

Notable present day examples of marginal seas with their access to the open ocean restricted by a swell are the Mediterranean (Gibraltar) and the Black Sea (Dardanelles). Marginal seas originating through graben tectonics are the present day Red Sea and, in the geological past, the upper Rhine Graben. In the region of the Red Sea a dry, hot climate has had a similar effect on both the hinterland and the marine basin (formation of desert, salt precipitation). In the case of the upper Rhine valley, however, the effect of a warm but only intermittently dry climate in the Oligocene was quite different in the hinterland (tropical vegetation) from that in the marine basin (evaporites). We can therefore suppose that a strong, previously concentrated liquor was introduced from forebasins (Paris Basin, Pfalz) into the tectonic trap of the upper Rhine Graben.

Other styles of tectonism, for instance salt tectonics, led to local facies differences in the neighbourhood of the rising salt body. Depressions often form on the margins of such diapirs (LOTZE, 1957; TRUSHEIM, 1957) and collect the detritus from the slopes. Not only do totally different raw materials in a narrowly limited area result in this way, but through reaction between the ground water and the salt a saline milieu can be created in a terrestrial facies. Leach depressions of this kind existed until a few decades ago in the region of Halle-Mansfeld (central Germany). They were connected with subterranean and partly leached salt pockets. These salt lakes differed from their fresh water neighbours in fauna (brackish-water ostracods, snails) as well as in chemistry.

The development of such depressions formed by salt solution need not, however, go so far as salt lakes. Extensive bogs and fresh water lakes can also result from normal salt solution along tectonic lines. In such depressions in the sub-Hercynian basin north of the Harz thick brown coal seams formed during progressive sinking in the Tertiary. The salt glaciers in the Zagros chain of southwest Iran are also manifestations of salt tectonics. They lead to originally salt-free zones being covered with salt débris, in later phases of their geological development under the influence of salt metamorphism, a totally distorted picture of the genesis of these zones will be produced.

A problem undoubtedly of no less importance for geochemical facies analysis is that of synsedimentary tectonics. In shallow sea areas (Lower Saxony Basin, Emsland) there may be a considerable effect on the distribution of sediment and therefore on the repeatability of geochemical values.

The great transgressions in the earth's history also had a tectonic origin. They spread over areas of the shelf margin and tableland (Russia) and work over the previously weathered beds. According to the already mentioned investigation by VINOGRADOV and RONOV (1956) certain elements produced by humid weathering are thereby enriched in the marine sediments. Almost world-wide transgressions of this kind occurred in the Cambrian, on the Jurassic–Triassic boundary, in the Neocomian, and on the Palaeocene–Eocene boundary. The transgression in the Saxony Basin (Lower Saxony) connected with the formation of iron ore deposits, is of more local importance.

Geochemical facies analyses, being concerned mainly with the sediments of peripheral seas and fore-deeps and only rarely with deep sea clays, is to a great extent dependent on tectonic influences.

PROVENANCE OF THE SEDIMENT

In the facies concepts of PUSTOVALOV (1940) and KLENOVA (1948) the origin of the sediment is also considered. It is a question which was undoubtedly neglected in the early geochemical facies analyses. The examples of the weathering of granite and the differing sedimentary assemblages in the fore-deeps have clearly shown in another connection how strongly geochemical facies investigations are dependent on the origin of the material.

It is first necessary to know from which source rock the various sediments and chemical elements could have been derived. Volatile compounds are usually enriched in hydrothermal and pneumato-lytic rocks. Granite and the majority of acid plutonic rocks contain mica and feldspar which on the one hand yield certain trace elements (As, Be, Ga, Ge, V, U) and on the other give rise on weathering and deposition to new clay minerals. Basic rocks are, in contrast, rich in Cr, Cu, Mo, Ti, and Sn.

This review is incomplete and is designed to show only that the weathering of granites can lead to a fundamentally different assemblage of trace elements and minerals as the decay of a basic rock. The details of the distribution of the various trace elements will be discussed in a later chapter of this book.

The situation with sedimentary rocks is similar: apart from Ca limestones yield essentially Ba, Fe, Mn, Sr, and very subordinate Mg, whilst clays contain the total gambit of trace elements. Metamorphic rocks are on average poorer in trace elements.

How does this picture appear when one compares the trace elements of a restricted depositional zone with those of the source area? Unfortunately few geochemical investigations have handled this problem in detail, owing mainly to the fact that by far the greater part of the minerals have been reworked and concentrated for the second or third time. A connection with the original source rock is therefore not easy to establish. A few detailed investigations have given the following results: in the catchment area of the Alpine Rhine in the vicinity of Lake Constance MÜLLER and HAHN (1964) were able to distinguish three carbonate and six heavy mineral distribution provinces. They indicate source areas in respectively granite and crystalline schist, flysch, Helvetic nappes, and middle and upper East Alpine nappes. Similar results were obtained by VON ENGELHARDT et al. (1953) for the area to the northeast of Lake Constance. BRANDENSTEIN et al. (1960) suggested that the trace elements in the coals and turfs of Austria were derived from the Bohemian Massif and the Alps. LEUTWEIN and RÖSLER (1956) were able to elucidate the erosion province from the trace elements in central German coals. The origin of the uranium in streams and sediments in Austria was investigated by HECHT (1963), and SAHAMA (1945) gave a review of the trace elements of Finnish rocks, finding that the Fe^{2+}/Mn ratio in the eruptive rocks was similar to that in the sediments (with the exception of black shales). In carbonate rocks he found Mn enriched in comparison to bivalent iron. The rare earths La, Ce, and Nd were associated with younger granites, granulites, greenstones and dolerites; in basic rocks they are present in considerably lower concentrations. Cr is a typical element of basic and ultrabasic rocks, and is much less rich in carbonates.

Further connections between the chemical make up of source areas and their sediments can be read from the balances drawn up for individual elements. These balances will later be considered in detail in the discussion of facies analysis. Most balances show that the trace elements in the sediment are in many cases determined by the composition of the source rock. Where the weathering of granite yields only kaolinitic material a high boron or chlorine content in the sediment is ruled out as a matter of course. Where, in contrast, marine clays outcrop in the hinterland a high original content of B and Cl is to be expected which is unlikely to be overprinted by facies effects in the area of deposition. The influence exerted on a limnic deposit by the chemical content of a marine sediment in its source area is shown by an example given by BAUMANN (1968) from the Tertiary foreland-molasse of southern

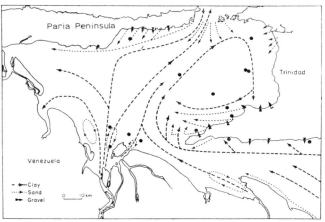

Fig.3. Directions of sediment transport in the Gulf of Paria. (From HIRST, 1962a, p.312, fig.2; with kind permission of the author and publisher.)

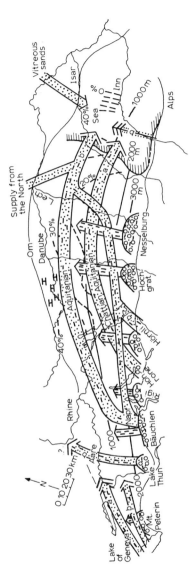

Fig.4. Sediment transport in the molasse basin north of the Bavarian Alps in the Late Oligocene and Early Miocene. (From FÜCHTBAUER, 1967, p.288, fig.11-3; with kind permission of the author and editor.)

Germany. Baumann encountered a high proportion of boron (average 0.03% B_2O_3) in a clearly limnic series of clay-marlstones (Gehrenberg Marl: upper fresh-water molasse). Investigation of the microfauna confirmed his suspicion that reworked Late Cretaceous marine material from the Helvetic zone of the Alps was present. Similar effects can occur if the marine floor of a limnic basin is eroded and carried into the centre of the basin (HARDER, 1963). Exact determinations of the origin of the sediment are, however, only possible in very few cases. They can be made if the constituent fossils yield information over redeposition of the sediment or give clear hints of the transport routes. How complicated the details of the sediment transport can be is shown in Fig.3 for the Orinoco delta (recent example) and Fig.4 for the south German Tertiary foreland-molasse (Chattian, Aquitanian, Cyrenea beds, lower fresh-water molasse).

RATES OF SEDIMENTATION

The original make-up of the rock determines to a considerable extent the chemical balance of the sediment, as the examples in the previous chapter show. To what degree and which chemical components the sediment can take up, however, is connected not only with the nature of the supplied material but also with the sedimentation rate and the composition of the sea water.

The sedimentation rate is dependent on the one hand from the climatic and tectonic events on the continents, and on the other from the palaeogeographic situation in the individual sedimentary basins. It must also be borne in mind that a comparison between the present day sedimentation rate and that of the past is very difficult because only a part of geologically old sediments are still preserved. GREGOR (1967), from whose papers all the following data about sedimentation rates have been taken, therefore divided the surviving sediment volume of a system by the number of years assigned to it on the geological time-scale. He arrived thereby at the "net annual rate of sedimentation". According to RONOV (1959)

$268 \cdot 10^6$ km^3 of sediment have been preserved from the $265 \cdot 10^6$ years from the Devonian to the Jurassic, which corresponds to an average net accumulation of 1 km^3/year. The deficiency in this calculation lies in the fact that folding and erosion are restricted to short periods of the earth's history and as a result the sedimentation

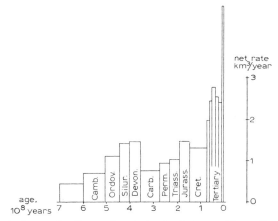

Fig.5. Survival rate of sediments in relation to their age. (From GREGOR, 1967, p.24, fig.6; with kind permission of the author and publisher.)

rate must also have experienced frequent changes (LIVINGTON, 1963). HOLMES (1965) therefore also considered the maximum thickness of the different systems since the post-Algonkian in his calculations of sediment rate. GREGOR (1967), too, based his "survival rate of sediments", shown in Fig.5, on this factor. This diagram clearly portrays the connection between mountain-building processes and sedimentation rate. As an example of the sedimentation rate in a specific area the Gulf of Paria (between Trinidad and Venezuela) described by HIRST (1962a) has been selected (Fig.6). The strongly varying sedimentation rates within a narrow area are very clearly to be seen. A further example has been taken from a paper by MÜLLER and FÖRSTNER (1968) and shows the fluctuations in the quantity of material in suspension delivered at the present day by rivers (Fig.7).

As a rule a high sedimentation rate means that the clay minerals

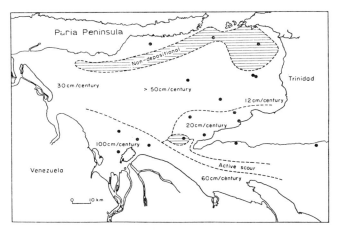

Fig.6. Recent sedimentatoin rates in the Gulf of Paria. (From HIRST, 1962, p.312, fig.3; with kind permission of the author and publisher.)

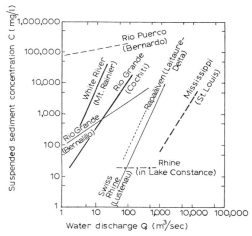

Fig.7. Suspended sedimentary material in various rivers. (From MÜLLER and FÖRSTNER, 1968, p.245, fig.3; with kind permission of the authors and editor.)

and organic complexes do not stay long enough in contact with the water to become saturated with its chemical components. For example HARDER (1963) reckoned that about 10^4 years were necessary for the complete incorporation of boron into mica minerals. When one remembers that during this time the mica-mineral need not be constantly in direct contact with the sea water solution then relatively long periods must elapse before saturation in certain trace elements is reached. This is also indicated by experiments with detritic chlorite which was treated with an $MgCl_2$ solution for 6 months. During this time JOHNS (1963) was unable to ascertain any significant increase in the Cl content of this mineral. In this experiment one cannot, of course, quite exclude the possibility that the detritic chlorite did not react in the same way as a newly formed chlorite mineral in respect to Cl take-up in the intermediate layers.

Nevertheless in most cases slower sedimentation leads to a higher incorporation quota of the trace elements. Low and high rates of sedimentation in a narrowly restricted region of deposition lead in any event to varying enrichment in trace elements, quite independent of the influence of facies. It is, of course, only the rate of incorporation in the clay mineral which is affected. Organic substances within the clay sediments, because of their large inner surface area, are probably capable of absorbing a notably greater quantity of trace elements than the clay minerals.

Further special conditions result from the precipitation of metal sulphides in an anaerobic milieu, which can lead to an increase in the trace element concentration.

GRAIN SIZE DISTRIBUTION

The majority of trace elements are bound to clay minerals, either in absorbed form or through direct incorporation in the lattice. In addition there are also links to heavy minerals (tourmaline, augite, hornblende) with a different grain size to clay minerals. The grain size of a sediment thus execises an influence on the distribution of trace elements. Of the most important facies elements the following are found on the coarse fraction: B (in tourmaline), F (in apatite, biotite, hornblende), Co, Se, S (in pyrite), Ge (according to HÖR-MANN, 1962) among others in zincblende), and Cr. In this coarse fraction practically only detrital and rarely newly formed clay minerals occur. Detrital and in some degree also newly formed clay minerals are, in contrast, restricted to the fraction smaller than 2 or 1μ. Here occur the chief concentrations of B (illite, muscovite, glauconite), Cl (chlorite), and F (illite/muscovite). Linked to organic substances are above all U, V, Cu, P, and according to GROSS (1964) also B.

These few examples are sufficient to show the importance of grain size for the distribution of trace elements. Therefore only trace element concentrations from sediments of identical grain size are strictly comparable with one another.

The breaking up of a rock into its size fractions is, however, attended by considerable difficulties if it is strongly consolidated. Thus the size fraction <2 μ of clay and siltstones from borings in the Bavarian Alp-foreland could only be separated up to a depth of 2,000 m, below this level the beds were too strongly lithified. Similar difficulties occur in using rock specimens of identical stratigraphy and facies, but one from a surface exposure and the other from a bore hole. In such cases usually only the specimen from the surface exposure can be broken up into its grain-size fractions. The only course remaining for dealing with the other sample is to test the unprepared rock as a whole for the sought after chemical components. In comparison with values from the clay fraction those from the total rock often differ by as much as an order of magnitude. For certain elements and rocks, however, the trends of the trace element content of the clay and heavy fraction are noticeably parallel so that a comparison between the contents of the various grain size fractions is quite possible. A large number of closely spaced total rock samples are, however, necessary. The question of the effect of grainsize on single elements in the establishment of ratios, e.g., V/Cr, is difficult. V is linked to the clay or the organic fraction, whereas Cr is enriched in coarser material (BAUMANN, 1968; GITTINGER, 1968). If one compares the V/Cr ratio of a silt with that of a clay-stone from the same hydrofacies, then the balance will be in favour of chromium in the coarser and vanadium in the finer fraction. The comparison in this case is between a dominantly hydrofacies-bound element (V) and another (Cr) with its concentration primarily controlled by the quantity of detritus. Therefore in this case also only the clay fraction should be employed; if strong lithification makes this impossible then a wide-ranging series of inseparable samples must be investigated.

Sandstones and conglomerates are the most unsuitable sediments for the diagnosis of geochemical facies: they contain practically no clay or organic substance and correspondingly exhibit only those trace elements which are linked above all with detritus. Moreover the pore cement of the sandstones is usually of a diagenetic nature and its chemistry may therefore be derived from substances present in beds at other levels. One can on no account turn to clay intercalations in the sandstone in order to determine its facies province on the basis of the chemical components in the clay. These layers can originate either from the reworked base sediment or from an environment which has nothing to do with the sandstone. In the first case we are commonly concerned with very thin 2–3 cm layers occuring several time one above the other in the sandstone.

General geological and palaeontological characters indicate that such sandstones may, for instance, be formed in a marine milieu close to a coast. BAUMANN's (1968) investigations have shown that in his example the same chemical components were present in the same concentrations in the clay intercalations as in the clay-marl base sediment of the sandstone. The clay layers were formed from reworked non-marine base sediment, and were insufficiently exposed to the marine milieu to take up its trace elements in significant quantity before the next influx of sand. In the second case the clay intercalations may have originated through marine incursions and be verified as such by Foraminifera in the clay-stone. The sandstones in this example are solely of fluviatile origin. Such cases occur several times in the youngest Late Carboniferous of northwest Germany.

BIOLOGICAL INFLUENCES

Biogenetic components also enter the various grain-size fractions and serve as a basis for other influences on the enrichment in trace elements. One must here distinguish between the finely ground shell fragments in the coarser fractions and the organic substance itself which occurs more finely distributed in the less coarse grained fractions. Both enrichment components enter into the analysis if no breaking up of the sample can be undertaken. Self-evidently the Ca content of a sediment is raised by the presence of shell fragments. In this manner the Mg/Ca or Sr/Ca ratios can be markedly shifted. Furthermore it must be remembered that between the shells of gastropods, hydrozoa, anthozoa, cephalopods, foraminifers and articulate brachiopods a significant gradation in the Sr content exists (THOMPSON and CHOW, 1956). Algae and corals, both with aragonite skeletons, yield high Sr values (ODUM, 1957). Apart from this differences in the Sr content may be governed by the growth rate of the shell (SWAN, 1956).

The shelly parts of the organisms contain, however, not only Ca, Mg, and Sr, but also B, Cu, Fe, and Mn in calcareous specimens (VINOGRADOV, 1953; PROKOFIEV, 1964; KREJCI-GRAF, 1966) and, especially in aragonite shells, Ba and Rb. These enrichments have been shown in recent investigations (LEUTWEIN, 1963) to be essentially dependent on the hydrofacies (especially salinity).

Through early diagenetic alteration of the shell material a proportion of the elements linked to the shell may reach the surrounding sediment. KULP et al. (1952) have established that for strontium the sediment (carbonate matrix) shows about half the Sr content of the shell. By stronger recrystallization this ratio shifts in favour of the carbonate matrix. On the other hand before the total decay of the prismatic layer the original shell material may, in a certain diagenetic milieu (O_2-poor), be replaced by secondary combinations (siderite, pyrites). This exchange brings new trace elements into the shell complex which have no immediate connection with the original hydrofacies. These processes have a special importance in the diagnosis of lithofacies (reef, off-shore mounds). The distribution of organic substance in fine grained sediments is also important. It is not quite true that these organic substances can only store secondary trace elements. The body fluids of these organisms and also the organic integument of plankton are rich in just those components of the sea which otherwise rarely enter the sediment (according to KREJCI-GRAF (1962a) apart from B and Cl, also Br and I). These elements remain in the sediment after the bacterial decay of the plankton like nitrogen and to some extent phosphorus in an oxygen free milieu. Of course some of the content of these substances is later given up with water on later compression of the clay.

WEEKS (1958) and HUNT (1962) reckoned that on average claystones have an organic content of 2.1%, carbonates have 0.29% and sandstones 0.05%. DEGENS (1968) calculated from these data that $3.8 \cdot 10^{15}$ tons of organic matter are present in sediments, of which $3.6 \cdot 10^{15}$ tons are in clay-stones alone (compare with the $3.6 \cdot 10^{12}$ tons in coal deposits). Unfortunately Degens's data do not reveal over which geological periods his calculations apply. Thus a considerable reserve of organic components which has considerable importance for changes during the diagenetic phase of sediment formation. Amongst these organic components are numerous complexes which are considered in geochemical facies diagnoses (e.g., amino acids, carbohydrates, hydrocarbons).

Plant life is a further important biological factor. Probably from early in the earth's history primitive lichens and moss were able to colonize cliff faces and to cause chemical changes in their immediate neighbourhood through photo-synthetic assimilation processes. As well as the ability to photosynthesize and the formation of gas during decay, plants have an importance in the bonding of trace elements. They withdraw important trace elements from the soil and concentrate them in the cell fabric so that after fossilisation they are enriched in the resulting deposit. Additionally, extensive swamp belts such as those present for example, during Late Carboniferous and Tertiary times, were able to soak up like sponges all the chemical solutions supplied to them. In this manner mangrove forests function today as coastal mud traps. Consequently coal swamps stored up a considerable number of trace elements.

INFLUENCE OF DIAGENESIS AND METAMORPHISM

The primary geochemical facies pattern can be overprinted by diagenesis and metamorphism. Diagenesis alters the mineral constituents and the composition of the pore water. In the course of diagenesis exchange of bases and alterations in the pore cement occur. The organic substance may also be affected to a considerable degree. According to H. H. Read and J. Watson (in MÜLLER, 1967) diagenesis can be defined as follows:

"Diagenesis comprises all those changes that take place in a sediment near the earth's surface at low temperature and pressure and without crustal movement being directly involved. It continues the history of the sediment immediately after its deposition and with increasing temperature and pressing it passes into metamorphism."

Thus in a certain sense the development of the sediment is continued by diagenesis, whereas in metamorphism such a strong alteration of the rock takes place that preservation of the original geochemical facies pattern can no longer be reckoned with. In the presence of the higher grades of metamorphism practically nothing remains of the primary or secondary, diagenetically produced minerals constituents. In these phases vigorous migration of ions and concentration even of trace elements themselves in totally new minerals also occurs. For example, in the epi-zonal metamorphic region the boron content of initially marine and brackish sediments is reduced by about half: from 0.040 to 0.018 in marine, and from 0.030 to 0.013% B_2O_3 in brackish water clay-stones (ERNST et al., 1958). In the deeper regions of the meso- and katazones of the Sau Alp (east Kärnten, Austria) the rocks generally have a B_2O_3 content of only 0.010% (ERNST and LODEMANN, 1965).

Therefore in the meso- and katazonal stages practically no geochemical facies analysis is feasible unless one only wishes to determine whether the rock is of magmatic or sedimentary origin.

The boundary between diagenesis and metamorphism is difficult to ascertain from the mineral constituents of the sediment alone since in both domains the same mineral associations may occur

(VON ENGELHARDT, 1967). A flexible boundary was accepted by CORRENS (1963), but he held the formation of mica and chlorite as characteristic of metamorphism.

For practical purposes diagenesis ends when the intercommunicating pore spaces are closed through physical or chemical processes. According to VON ENGELHARDT (1967) reactions in this zone begin in the solid state or through diffusion along the grain boundaries of the minerals.

In these circumstances only the diagenetic influences on the original facies pattern can be taken into consideration. In the opinion of a number of authoritative authors (FÜCHTBAUER, 1956; MÜLLER, 1967; VON ENGELHARDT, 1967) diagenesis begins immediately after deposition of the sediment. MÜLLER (1967) separated the processes between deposition and early diagenesis and for the marine environment designated them as halmyrolysis (in the sense of HUMMEL, 1922; PACHAM and CROOK, 1960) and for the fresh water environment as aquatolysis (this expression would be best used as a superior unit for both environments since water solution [= aqualysis] is equally applicable to both marine and non-marine sediments).

For the individual stages of diagenesis MÜLLER (1967) introduced the following terms: (1) pre-burial stage; (2) shallow burial stage; and (3) deep burial stage. In the first stages of diagenesis the minerals of the sediment are still in contact with unchanged sea water. Hence the incorporation or adsorption of trace elements can continue. According to CORRENS (1963) the transformation and building of new minerals may begin here since abundant Mg is available in the sea water as well as Al and Si in the sedimentary material. The formation of tri-octahedral chlorites and their mixed-layer-minerals probably takes place during this stage of diagenesis (GRIM and JOHNS, 1954; POWERS, 1954; PINSAK and MURRAY, 1960; CORRENS, 1963). The formation of early diagenetic illite is, on the other hand, not certain. GRIM et al. (1949) described a new formation from degraded illites. GRIM and LOUGHNAN (1962) ascertained the building of illite from degraded vermiculite (14 Å) in a marine milieu. The transition from montmorillonite to chlorite or to sudoite and illite has been proved by experiment (WHITEHOUSE and MCCARTER, 1958). Using a starting material of suspended montmorillonite and artificial sea water only three years were necessary for the change over. However, a large range of new formations follows only in the later phases of diagenesis.

The formation of talc in anhydritic rocks of the Zechstein sea probably took place in the early period of diagenesis (FÜCHT-BAUER and GOLDSCHMIDT, 1959). The same authors were of the opinion that serpentine, pyrophylite, montmorillonite and corrensite could also be formed during this stage. These new mineral formations are all tied to the super saline environment where the Mg content of the sea water is especially high. The formation of new feldspar minerals has also been observed in the same milieu (FÜCHTBAUER, 1956; FÜCHTBAUER and GOLDSCHMIDT, 1959).

A new minerals formation often investigated is glauconite. The chief difference from illite consists, according to GRIM et al. (1949) and MÜLLER (1967), in a higher ratio of "di- to trivalent iron" in glauconite, as well as the higher total iron content at the expense of Al in the octahedral layer. Following the investigations of BURST (1958a, b), MILLOT (1964), and FAIRBRIDGE (1967) glauconite must be seen as a product of the pre-burial stage (MÜLLER, 1967). The high boron content of glauconite (according to Harder several thousand p.p.m.) is in this stage extracted from the pore water, which to this point probably still has the composition of the sea water (CORRENS, 1963). We can therefore reckon with similar processes during the incorporation of boron in detritic illite clay minerals, and therefrom can deduce that the incorporation of special trace elements can follow in water of the composition of the sea water even under a cover of further sediment. The real geochemical facies pattern originates in such cases only after the deposition of the sediment and not, as commonly assumed, during deposition.

The reduction of pore space also plays a role in diagenesis. In clay sediments, which are decisive for geochemical facies analysis, the average porosity of freshly deposited sediment becomes reduced from 0.8 to 0.70–0.65 at a depth of about 7 m (VON ENGELHARDT,

1960). The first comprehensive investigation of the porosity of recent and older clay sediments can be traced back to HEDBERG (1926), who was able to show a regular decrease in porosity with depth. This compaction of clays, and to a lesser extent sandstones, has in the meantime been confirmed by many authors. According to several corroborative investigations the pore volume of clay amounts to only 0.10 at 2,000 m ($^{1}/_{8}$ of that in fresh sediment). A reduction of the water content of the clay is linked with the reduction of pore space. A change in the salt content, pH, redox potential and the temperature of the remaining pore solution may also result. In the evaluation of geochemical facies analyses it is important to know approximately when and at what depth changes in the pore water can occur.

In a marine basin of south California EMERY and RITTENBERG (1952) found an increase in the pH from 7.59 in the upper layers of the sediment (sea-floor water 7.52) to about 8.5 at a depth of 8 m. This pH increase with depth could not be confirmed in other recent sediments of the Atlantic and Pacific (SIEVER et al., 1965). There the pH fluctuates between 7.2 and 7.7. However, the pH values in the sediment below the pH-table of the sea water sank as soon as bacterial decomposition of the organic substances caused the partial pressure of CO_2 to rise beyond the CO_2 pressure of the atmosphere. According to MÜLLER (1967) this affect probably occurs 10 m below the surface of the sediment.

Further changes in the pore water occur through sulphate reduction. These changes are also brought about by bacteria which can already be active in the upper centimetres of oxygen-free sediments. The pore water correspondingly decreases in SO_4^{2-} with depth. The origin of "hydrotrilite" (FeS $\cdot nH_2O$), which gives the sediment its black appearance, is connected with the production of H_2S. In gyttja the boundary between the light coloured oxidation zone and the black reduction zone lies a few centimetres to decimetres deep in the sediment.

The earliest change in the pore water is thus also influenced by the quantity and distribution of organic material. In the oxidising milieu of purely mineral beds the commencement and depth of the chemical transformations are displaced in comparison with bituminous sediments. We can therefore only recognize an average depth in the sediment at which the composition of the pore and sea water is still similar. It probably lies at less than 2–3 m, so long as sapropelitic conditions are not present. Assuming an average rate of sedimentation of 20 cm/100 years, as is the case in present day clay sediments in current-poor areas (e.g., Gulf of Paria, Venezuela), then the clay minerals remain in contact with the sea water for about 1,000–1,500 years. In the changes which then follow the SO_4^{2-} content and calcium are first affected; the latter remaining in solution or precipitating out, depending on the CO_2 balance. Further changes are connected with the exchange of bases, whereby the pore solutions become impoverished in Na and gain exchanged Ca. Clays strongly absorb K which is also used in the formation of secondary micas. According to VON ENGELHARDT (1960, 1967) not only does the formation of new minerals occur but also the solution of already existing minerals. In this manner, through the solution of feldspars in a somewhat later stage of diagenesis, silica and aluminium as well as alkalis and calcium originate. The production of kaolinite in sandstones is connected with the solution of silica.

Into the pore water are brought, in addition, the sea water components in the body fluid of plankton (I and Br) and those derived from fossil hard parts (see chapter 3, section on biological influences). The chemical decay of the animal soft parts occurs during, and especially after, the solution of the external skeleton. In the upper, well aerated, zones of fresh sediment the unstable proteins are almost totally destroyed. In the boundary area between gyttja and sapropel the proteins and fat form the so-called body wax (WASMUND, 1935), which consists of free fatty acids and their NH_4^-, Na-, and Ca-soaps. Skin, muscle, lungs and brain may remain preserved in this manner. Organic structures are also often recognizable and complete skeletons may also be found in this boundary area of the redox-potential. In addition the preservation of chitin and skleroprotein through encrustation by lime is also favoured.

During deposition and early diagenesis in the totally oxygen-free milieu of the sediment only the body wax is somewhat structurally

rearranged, but even unstable substances such as protein and gly-
cerine remain chemically unchanged (the animal raw material
travelling several times through the body of the bacteria according
to KREJCI-GRAF, 1963a).

Information over the fate of the porphyrins is plentiful thanks
to the work of TREIBS (1934), BLUMER (1950), GROENNINGS (1953),
and DUNNING et al. (1954) as well as others. The greater part of the
porphyrins probably originates from chlorophyll; a further part is
probably linked with haemin. Whilst the transformation of chlorophyll in oxygen-bearing water proceeds relatively rapidly, it is
preserved in gyttja as well as sapropelic zones. In this connection the
combination of metals (V, Ni, and to some degree also U) with the
porphyrins is important.

The decomposition products of the organic substances are at first
held in the immediate neighbourhood of the buried fossils. They
often form there enrichment zones (HELLER, 1965; DEGENS, 1968).
In these enrichment zones the diagenesis of the clay minerals and
the changes in the pore water differ from in the non-encrusting
sediment zones.

The preservation of transformation of the carbohydrates in sediment depends on somewhat different factors. The commonest carbohydrates occurring in sediments may be based either on a simple
sugar (monosaccharide) or a polysaccharide. The free sugars are
hardly attacked by weak acids in the soil whereas oligo- and polysaccharides are depolymerized step by step. In the case of cellulose
the same authors wrote that strong acids were necessary in order to
efficiently break down glycoside bonds between the individual mo-
nomeres. The remarkable preservability of cellulose is, however, explained by the fact that oxygen is more quickly consumed by humic
water, e.g., in moors, so that even the less stable cellulose remains
preserved (KREJCI-GRAF, 1966).

Free sugar (common monosaccharides and also raffinose and
maltose of the oligosaccharides) is still found in sediments 10,000
years old. As a rule, however, the quantity of free sugar sinks to nil
or a very few % within a few metres because of bacterial action
(DEGENS, 1968). Combined sugar, e.g., cellulose, has been isolated

from coal of Early Tertiary age. According to DEGENS et al. (1963) and RITTENBERG et al. (1963) in marine sediments the content of sugar becomes reduced in the following manner: surface of the marine sediment, 500–3,000; 0.03–0.05 m, 100; 10–150 m, 10–20 μg/g dry weight.

An indication of the reduction of organic substances in defined organisms has been given by ^{14}C dating of shells of *Mytilus californianus*. Apparently a clear reduction of the organic shell substances in comparison with recent animals occurs within 400–5,500 years (HARE, 1962). From these data regarding the transformation of organic and inorganic substances during early diagenesis we can at this point draw the following preparatory conclusions:

(*1*) With the exception of the super-saline zone the formation of new clay minerals is rare and probably restricted to chlorite.

(*2*) The break down of organic substances in the sediment leads, amongst other things, to the formation of CO_2 and therefore to differentiation in the solution and precipitation of Ca (Mg).

(*3*) Through the exchange of bases and concentration changes during the compaction of the clay Na, Mg, and K are lost to the pore water.

(*4*) Changes in the pore water do not affect Cl and the majority of trace elements (B, Br, I); SO_4, however, is at first strongly reduced with depth through sulphate reduction.

The factors listed under (*4*) make it likely that B and Cl are present in sediment for a relatively long period in about the same order of magnitude as in the sea water. This period is probably roughly the same as that necessary for the total saturation of detritic illite in boron. The later formed illite is probably able to extract ample boron from the pore water since the reconstruction of montmorillonite into illite takes place in MÜLLER's (1967) opinion in the pre-burial to early shallow burial stage, and according to CORRENS (1963) extends at least to the shallow burial stage.

In certain circumstances the NaCl concentration may also be preserved to considerable depths. This has been indicated by SCHUSTER (1963) who, with the aid of electrical bore-hole measurements in a depth of 2,000–4,000 m in northwest Germany, was able to recognize quite convincingly several marine horizons in the Late Carboniferous on the basis of high Na/Cl ratios. On the whole, however, the existence of such connate water is strongly disputed. From the investigations of both KREJCI-GRAF (1960) and VON ENGELHARDT (1960) it is clear that the Na/Ca ratio and the Cl concentration in formation waters suffer marked changes which are practically only connected with the petrographic composition, tectonic position and depth of the sediment. In oil field waters the electrolyte content may rise to ten times that of sea water and recent sediments. In comparison with sea water the pore water in deep burial zones shows a decrease in the Na/Cl, K/Cl, Mg/Cl, HCO$_3$/Cl, and SO$_4$/Cl ratios whilst Ca/Cl increases (MÜLLER, 1967). These changes affect above all the pore water of clay sediments and in the opinion of DEGENS and CHILINGAR (1967) are based on ionic filtration by charged-net clay membranes.

The creation and migration of formation waters have still not been satisfactorily explained. Since the development of formation waters begins practically with a strong contraction of the clay and the complete transformation of the organic substances, the chemical facies pattern is not greatly influenced. The transformation of minerals in the last phase of diagenesis (deep burial stage) concerns most of all the kaolinite which with increasing depth and decreasing porosity is changed into chlorite (ECKARDT, 1958; FÜCHTBAUER and GOLDSCHMIDT, 1963) or rebuilt as sudoite (VON ENGELHARDT et al., 1962; KROMER, 1963).

With the approach of the boundary between diagenesis and metamorphism lies a relatively monotonous illite–chlorite paragenesis, which may originate from various starting materials (montmorillonite, kaolinite, in part also illite). During low temperature metamorphism the proportion of chlorite grows whilst illite is transformed into sericite. These tendencies in the changes in pore water and minerals during the course of diagenesis should be specially observed in geochemical facies analysis.

The course of the diagenesis of coals is much clearer. It is characterized by five stages: bog—peat—soft brown-coal—hard brown-coal. In the opinion of many coal petrographers actual bituminous coal is a product of weak metamorphism. In this step in the diagenesis changes occur connected with the redox-potential just as in inorganic sediments. Thus in the oxygen zone intensive decomposition of the vegetable matter takes place. However, the quicker the bog sinks under the water table or is covered by clastic sediment, the more of this original material is preserved; it only suffers structural rebuilding by bacteria in the anaerobic zone. Bog, peat and soft brown-coal are therefore characterized by biochemical transformations, whereas during the transition to hard brown-coal physical processes dominate: the pores of the coal are substantially squeezed together and the pore water expelled.

The diagenesis of the coal and its seat-earth or roof-rock need not proceed synchronously. An example from the Late Carboniferous of the Saar region shows that the coal there was sufficiently compact for vertical fractures to form, whilst in the clay-stone bed above considerably more pore space was preserved and fractures were not formed. Along the vertical fractures Ca, Mn and Fe were extensively dissolved, carried away and finally enriched in the top of the coal and in the succeeding clay-stone. The enrichment zone in the clay-stone penetrates some two metres into the roof.[1] This distribution can only be understood if Ca, Mn and Fe have been dissolved under the agency of CO_2 and then transported along the fractures into the roof-bed of the coal. Here, as a result of decreasing CO_2 pressure in the relatively loose rock fabric, they were deposited as carbonate. The diagenetic state of the coal at this time can be said to be in the weak brown-coal state with regard to the expulsion of CO_2 and water, whilst the formation of vertical fractures is indicative of the hard brown-coal stage. The fractures do not continue into the surrounding rock, and we must therefore have to deal with flow processes which set the whole contents of the pores under movement. Permeability is therefore necessary and hints activity in an even earlier stage in the diagenesis of the clay-stone. DEGENS (1958) referred to migration of material in the seat-earths and roof-beds in coal-yielding sequences.

[1] Report by A. Baumann and K. Gittinger in contract to the Saarbergwerke AG., Saarbrücken (1967).

We must therefore bear two points in mind when considering beds with thick vegetable intercalations: (*1*) the diagenesis of the clastic and organic components does not proceed synchronously; and (*2*) in the roof-rock above the coal, chemical transfers may occur which, if they involve easily soluble elements, can lead to an overprinting of the geochemical facies pattern.

CHAPTER 4

Methodological Prerequisites

INTRODUCTION

For the determination of geochemical facies there are a series of methodological pre-conditions. In part they result from the dependence of the facies on geological and mineralogical factors, and concern the taking of samples and their number (p.47); the problems of apparatus used in chemically preparing the samples and in their analysis (p.52); the gauging of the results of the analyses with geological, palaeontological and mineralogical criteria (p.55); and the incorporation of the results in an overall picture of the investigated sediment (p.60).

The heavy dependence of the geochemical facies on the factors discussed in chapter 3 necessitates meticulous sampling and the investigation of as large a number of samples as possible. Only in this way is a statistical evaluation over lithological units and different outcrops possible, and in this way a number of possible errors cancel each other out. The other method consists of as meticulous an examination as possible of the element distribution with very many individual characters of the rock and in this way also obtain reliable information about linkage and dependence of the elements. However, these investigations are valid solely for the locality examined and can only be carried over into a large depositional area with considerable reservation.

The time consumption is probably roughly the same for both methods of geochemical facies investigation. The appropriate method must be chosen bearing in mind whether the main aim is to discover the correlation between the chemical elements and the minerals, or the distribution of the elements in relation to certain recognizable depositional features. Since the facies can commonly change within a small area the statistical method of examination over a greater thickness of beds and area of exposure is to be recommended. That does not, however, exclude a more exact investigation of one or more standard profiles over lesser lithostratigraphical intervals.

With the large number of necessary samples appear concomitant analytical problems. The present day spectrographic apparatus as well as those for X-ray fluorescence and microprobe investigations permit simultaneous analysis for several elements with comparative accuracy. The time consumed by the analysis is correspondingly small, and this also applies to the preparation of the samples for analysis. In multiple analysis, however, difficulty arises in the standardization of the samples for each analysed element: a comparative standard of the same order of concentration as in the sample itself is necessary. When a number of samples are processed the standard of a certain element suitable for one may well be inapplicable to one or more of the remainder.

The other possibility in whole rock analysis is to restrict the investigation to a few elements which appear to be diagnostic for the facies zones. In this case it is possible to use standard processes which yield a rapid flow of analyses. Such methods are based on atomic absorption spectrography, spectral photometry, and flame photometry as well as gravimetric and volumetric analysis. The chemical elucidation of the samples then requires substantially more time than other processes. The handling time is also increased by the fact that the individual elements cannot be simultaneously determined. On the other hand the standardization of the elements is both simpler and more accurate since only one or a few standard elements are involved. The advantages and disadvantages of these several methods will be discussed individually in the following chapters.

SAMPLING

In taking samples one must always consider whether the facies alters within a narrow area or whether fluctuations are present only on a large scale. In areas with attenuated sediments (a few centimetres or metres representing a chronological unit expressed elsewhere by substantially thicker beds) sampling must be carried out at centimetre intervals. If this is not done one may easily include the sediment of several facies in the one sample. In thick limestone sequences without a visible character change, or in monotonous red-bed deposits, the sampling distance can, however, be considerably greater. In both cases one must ascertain the situation prior to sampling through a geological reconnaissance. In addition it is wise before beginning a large scale investigation to select a small standard profile and investigate it for as many elements as possible at small intervals. In this way information over the expected order of magnitude of the elements can be obtained for the chemical and technical investigations, and moreover precautions against disruptive chemical combinations can be taken.

As a rule orientation samples can be taken at a distance of 0.4 m,

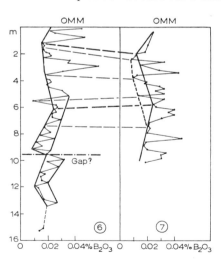

Fig.8. Correlation of the boron content of various profiles in the lower fresh water molasse (Tertiary/Aquitanian) on the boundary to the upper marine molasse (unfaulted molasse northwest of Lake Constance, southern Germany). (After BAUMANN, 1968.)

and in specially interesting profiles at half this distance. For exposures which are only temporarily accessible (working quarries and mines, borings, tunnels and impermanent escarpments laid free by road works) a smaller sampling interval should be chosen from the start. For an examination of the lower fresh-water molasse in the German Lake Constance area a sample distance alternating between 0.4 and 0.2 m was selected. Fig.8 shows how this alternating sampling distance is diagrammatically reflected in the results. A smaller sample distance was used for the Keuper marl in the vicinity of Tübingen since the samples were taken from trenches which were later to be filled in (Fig.9). Close samples should also be taken from above coal seams in order to recognize the diagenetically caused concentration changes in the carbonate.

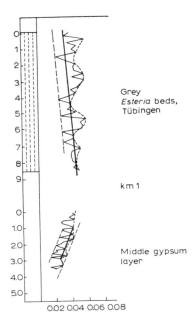

Fig.9. Increase in the boron content (minimal, average, and maximal values) in the grey *Esteria* beds of the Middle Keuper on approaching the middle gypsum layer (Pfäffingen and Tübingen, southwestern Germany). (From ERNST, 1966; with kind permission of the editor.)

The same provisions apply to rocks with rapid petrographic transitions.

Through as narrow a sample distance as possible extreme results caused by irrelevant properties of the rock are included in a statistically valid number. This is important not only for standardization but also for the extrapolation of the results to the remainder of the depositional province (see sections on environmental and geological evaluation of geochemical facies analysis).

The maintenance of the same sampling distance should be an essential condition for geochemical comparisons. Unfortunately, however, the same distance often cannot be adhered to from exposure to exposure. Thus fragments for investigation from bore holes are often only available from specially important stratigraphical or petrographical horizons. In mines the considerable accumulation of loose material may make regular sampling difficult. But even above ground the fluctuating outcrop conditions often prohibit continuous sampling. Therefore in a broad investigation area the most favourable long profile, be it a boring or road cutting or whatever, should be fully exploited and samples taken at narrow intervals. From the chemical analysis of these samples characteristic orders of magnitude and trends can be obtained from smaller and less well exposed profiles. Fig.10 shows individual partial profiles from the lower fresh-water molasse compared with a continuous profile from a deep boring. The investigation of very thin beds or others of uncertain stratigraphy is scarcely useful. For instance a 1-m profile is barely sufficient to provide clear evidence of the facies.

The individual sampling points should be exactly measured since diagenetic or petrographic changes in the rock may occur from centimetre to centimetre. For this reason it is also to be recommended that the sample itself should be taken from as precise and small an area as possible and not, for instance, across the bedding or along the length of the profile. Thus in a 20-cm column of rock, perhaps from a bore-hole, ground up and divided for analysis into four, the danger of a diagenetic or petrographic change is four times greater than if a 5-cm piece had been broken off and used.

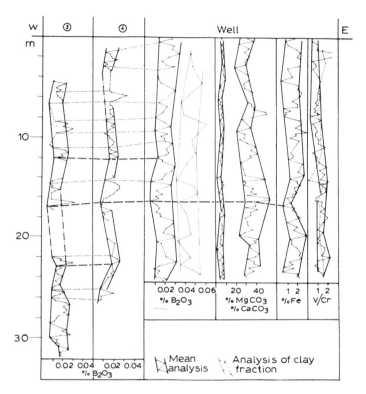

Fig.10. Examples of the correlation of the boron content for different profiles of the lower fresh water molasse (USM) northwest of Lake Constance (marginal facies of the unfaulted molasse of the northern fore-Alpine basin) and the basin sediments of the same beds in a well at Tettnang (Lake Constance, southwestern Germany). (After BAUMAN, 1968.)

The number of samples depends on the character of the investigation and can be arranged to suite it. If a stratigraphically limited unit is to be examined it is clearly possible to take geographically wide-ranging comparative samples which would lead to an impracticable amount of material in the case of a thicker unit.

The amount of material in the sample depends on the proposed investigation. Nowadays in the examination for trace elements a few grammes of rock are sufficient, and this is also the case for the determination of Ca, Mg, and other carbonates. Simply from the point of view of transport and storage samples, should be no bigger than to fit comfortably in the palm of the hand (approximately 100 g). What material is not used for the analysis should in all cases be kept for repetition and checking.

The packing of the samples depends also on the type of investigation planned. Analysis of the inorganic components requires no special measures in this respect. Bags of strong paper, easily labelled, are sufficient. Very wet samples or those with a tendency to decay are, however, best collected and stored in plastic (polyester) bags. This applies also to samples which after taking must be stored air-tight, it being possible to hermetically seal plastic bags with special clips. If total exclusion of air is necessary then brass cylinders with a sealed, screw lid and two needle valves can be used. After placing the sample in the container the air can be driven out with argon. This procedure necessitates only small pressure bottles of argon. Analyses of special organic components require caution in the choice of the plastic bag. Polyester and related substances can change with a raising of the temperature and humidity (through the sweating of the sample). The designation of the outcrop and the sample number must be quite clear from the labelling of the sample. All other sample specifications (geological formation and age, sample distance, outcrop description) are best entered in a special sample book.

The description of the sample must be made soon after it has been taken since after lengthy drying out and air storage the colours of the rock may fade and decomposition of the sample may set in. As a rule the description comprises all the external characters important for the analysis and evaluation of the sample. To these can be reckoned rock-type, grain size of sandstones, silt content of clay-stones, limestone content, solidity, inclusions (pyrites, siderite nodules), tectonic characters (dip, joints, slickensides), and fossils.

PREPARATION AND ANALYSIS

Before chemical analysis the samples are sorted according to rock-type and externally recognizable characters. In addition samples from bore holes must be cleaned of lubricant traces which can otherwise be a source of error. The samples air dried, or otherwise dried at 105°C, are broken with a hammer or a press, milled to a size suitable for analysis, and finally divided into one part for the total analysis and another for the preparation of grain-size fractions. In the separation of grain sizes the clay fraction (less than 2μ) is important above all others. The details of the various methods of separation are dealt with in numerous specialist books.

The further processing of the prepared sample depends on the method of chemical examination chosen. Therefore no binding directions for particular methods of investigation can be given here. The most essential differences between the more important investigatory processes have already been mentioned in the introduction to chapter 4. The following remarks only touch on general questions of method, which can be of some importance for geochemical facies analysis.

Before the preparation of new solutions for use in gravimetric determinations and titrations the raw chemicals should be examined for the element under investigation. Even guaranteed pure chemicals may occasionally exceed the boundaries of tolerance. These raw solutions must be stored in such containers that do not react with the solutions they contain. Therefore synthetic or quartz glass vessels are recommended for boron investigations since normal glass always contains boron. Mixtures of different compounds should not be stored for lengthy periods since they may easily tend to unmix.

Very decisive for the precision of a geochemical investigation is the chemical disintegration method. The choice of a method in gravimetric, spectrophotometric, flamephotometric and also atomic absorption spectrometry is dependent on whether silica and aluminium are to be looked for or only special trace elements. In the first case a hydrofluoric disintegration is necessary whilst for the second a soda disintegration is often sufficient. Before beginning a series of investigations of both methods should be tried on the same sample material in order to be able to apply the best right from the beginning. It is also advisable to test at the beginning various melting temperatures and times. Dependent on these to a certain extent are the texture of the melt and the solubility of the melt-cake. In just such investigation series it is important that the melt should be easily soluble so that the platinum crucibles are quickly reavailable for further disintegrations.

If it is possible that the trace element to be determined is linked both with clay minerals and organic material then different disintegration methods should be used. With suitable solution agents extractions can be made in order to draw off soluble finely distributed organic substances, which can be treated separately. Vegetable and carbonaceous concentrations in the sediment should be separately handled with a suitable solution agent and a further portion of the sample burned. Treatment with an ultra-sonic apparatus together with the applicable solution agent is useful in the extraction of the finely divided organic material (MÜLLER, 1964).

Analysis of amino acids, carbohydrates, and fats in the sediment requires special care. In this case caution must be applied to such an extent that the samples must be kept under-cooled and air-tight until analysis, in order to avoid any bacterial decomposition of these components. In addition the body fluids of the bacteria would strongly falsify parts of the analysis. Both BAJOR (1960) and DEGENS and BAJOR (1960) deal with the details of such investigations.

Sub-surface geochemical facies investigations require a special strategy. An important case is the areal restriction of black shales which are distributed not far below the surface, and are not determinable with normal geological mapping methods.

The close limitation of rocks of the oxygen facies can be made with the help of formation-gas (ERNST, 1968). For this purpose bore holes were sunk for a metre at distances of 20–40 m, and after the expulsion of air from the vicinity of the bore hole were left to stand closed for a time. After a waiting period methane and carbon dioxide are measured directly on the bore hole. Up to the present it was usual to store the captured gas in glass sample containers until analysis. However, standard measurements with a known quantity of test gas have shown that diffusion of the light constituents of the gas can occur through the in- and outlet taps. Within $2^1/_2$ h, 80% of the methane was lost. Measurements of the hydrocarbon gases should therefore be made directly on the bore hole whenever possible.

By all chemical analyses it should be remembered that series determinations over a period of time are always more certain when the same apparatus, chemicals, and laboratory assistants are used. Each change of equipment or personel have its corresponding effect on the analyses. Repetition of the analyses is to be recommended in the case of discrepant values and after a large number of treated samples.

The concentration data of the determined elements or compounds can be represented in different ways. It is customary to give weights from gravimetric analyses as percentages and from spectrographic tests as parts per million (p.p.m.) or grammes per ton (g/t). In the international literature p.p.m. is now most usually employed. In giving percentual weight data the number of places after the decimal point should not exceed the experimental accuracy of the analysis. The boundaries of error of the individual methods must be carefully worked out and especially in geochemical analyses, where one is in any case relying on very small differences in weight, should be rounded-off upwards rather than downwards. Methods of which the margin of error lies in the order of magnitude indicative of a facies difference are best not employed in geochemical facies analysis.

ENVIRONMENTAL EVALUTION OF GEOCHEMICAL ANALYSIS DATA

A number of geochemical facies investigations led in the past only to uncertain and very controversial results. The opportunities for relating the geochemical information to standard real environments were not fully exploited. What possibilities in fact exist for making such correlations and with what accuracy can they be applied?

Clearly the recognition and definition of the present day facies must lie at the base of any attempt at standardization of ancient environments, and the information on which they are recognized must be traceable in the lithified sediments. However, the facies definitions discussed in chapter 2 scarcely allow extrapolation between the geochemical environments of the past and the present day: most of them, as was pointed out, ignore geochemical information altogether. Pustovalov's definition is applicable but not sufficiently refined. He dealt solely with the marine domain and its division on the basis of redox potential into oxidising (siderite) and reducing (hydrosulphide) facies. Moreover, these two environments are not, in fact, restricted to marine conditions and may occur in fresh-water lakes on the continent. It is even impossible to make the assumption that because of the influx of weathering solutions into the sea marine sediments contain higher element concentrations than those of the continent. Thus Ga appears to be richer in continental than in marine beds (DEGENS et al., 1957, 1958; contrary to the opinion, however, of LANDERGREN and MANHEIM, 1963). Similarly there are strong continental concentrations of boron, for example in the borax lakes of North America.

Thus standardization of ancient, geochemically determined facies relies on the refined definition of present environments, and the recognition of unequivocal indicators for these environments common both to the present and to the past. One of the best of such indicators are fossils with extant relations from which ecological information can be drawn. Benthonic forms are best since they are normally unable to escape from unfavourable changes in the environment. Nevertheless both KREJCI-GRAF (1963b) and REMANE (1963) have made clear that not all groups are suitable since some display a tolerance outside the limits of the environments we normally wish to recognize. Thus of the echinoderms *Asterias rubens* can tolerate a reduction in salinity to 8‰ which is less than a quarter of the salinity of the open oceans. Similarly *Balanus improvisus* and *Neomysis vulgaris* according to REMANE (1963) can both withstand local freshening of the sea water. In the opposite fashion certain limnic forms (e.g., *Limnaeus (Radix) ovata; Theodoxus fluviatalis*) can survive in salt water up to 15‰. MÄDLER

(1963) has demonstrated that the Mesozoic Characea lived in brackish water, whilst the present day species occur mainly in fresh water. The *Dreissenidae* and *Limnocardiidae* migrated in the Tertiary from the Paratethys into fresh and brackish water (KREJCI-GRAF, 1963b; PAPP, 1963). Approximately simultaneously fresh water gastropods in the same area became adapted to brackish conditions and developed a wide range of forms.

The judgement of brackish water conditions is especially difficult. Forms belonging to genera found also in the sea but rarely in fresh water are dominant in this environment. According to REMANE (1963) the only exception is in the Pontocaspian region where brackish water forms are present in great numbers and in exclusive genera. The brackish water fossils are distinguishable from their marine relations through size, shell thickness and external characters. Moreover they usually build a specifically poor but individually rich fauna.

Certain identification of fresh and brackish water formations in the geological past is by no means easy. In the Late Carboniferous, for instance, *Leaia* occurs on the one hand with agglutinated Foraminifera and on the other hand with fresh-water fossils in limnic coal basins. Ostracods (*Jonesina* group) can occur both with *Lingula* and non-marine pelecypods.

Often the association of different facies fossils can be caused by water currents and wind. Limnic molluscs may be carried for many kilometres into the sea by rivers (according to Remane the transport distance is increased by the presence of air or gas bubbles on the spiral of molluscs). Brackish-water snails, e.g., *Hydrobia ventrosa*, can be found in an inland salt lake (compare chapter 3, section on tectonic influences). In the brackish Baltic Sea Foraminifera can be found close to the inflow of rivers (ROTTGARDT, 1952), and *Vivipara* may be associated with *Macoma, Cardium* and *Mytilus* far from the influence of any river.

However, there are some stenohaline (facies constant) organisms which live either only in marine or fresh water. To these belong in the marine regime: Radiolaria, corals, calcareous sponges and cephalopods, and brown and green algae. Recent diatoms can in general be associated with either fresh or brackish water. From the limnic facies zone fresh-water snails, certain species of diatom, as well as higher vertebrates must be mentioned as reliable facies indicators. Of the present-day higher plants practically all live on land; only mangroves, sea-grass (*Zostera*) and the moss *Fontinalis* can tolerate salinity (REMANE, 1963). Of course plant remains (including pollen) can be transported far out to the sea by rivers and wind. In periods of intensive plant development we therefore find numerous plant intercalations in marine beds, as for example in the Late Carboniferous.

From this discussion it is obvious that for the standard comparison of geochemical data not only individual facies fossils but also whole fossil associations should be considered.

Comprising several genera such a unit is less likely to have changed its environment than an individual species, but clearly its authenticity as a natural association and not an assemblage artificially swept together from different salinity zones must be verified. Fig.11

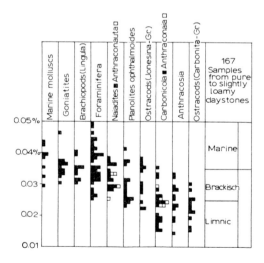

Fig.11. Standardization of the boron content of claystones based on the facies character of the contained fossils. (From ERNST, 1963, p.256, fig.1; with kind permission of the editor.)

repeats and earlier attempt made by the author to standardize the boron content of Late Carboniferous clay-stones against the fossil evidence. Marine pelecypods and goniatites, and limnic pelecypods and ostracods allowed the characteristic boron content of these two environments to be determined. However, the sediments of the intervening brackish-water milieu, palaeontologically somewhat dubiously characterized by non-marine pelecypods together with Foraminifera and burrows, could only be delimited with considerable uncertainty.

Palaeontological standardization is not always possible. Indigenous fossils (i.e., not wind blown spores) are primarily absent from super-saline deposits, and are often destroyed in beds which have suffered diagenesis. Even in normal sequences, e.g., foredeeps, there are often no suitable fossils. The greater part of the German Rotliegendes (Early Permian), Buntsandstein, and Keuper is almost devoid of environmentally indicative fossils. In such cases sedimentary structures (fluviatile accretory bodies, dunes, and glacial moraines on the one hand; barrier beaches, gullies, tidal-flat bedding, current structures etc. on the other), can be used as indices of the terrestrial or marine facies domains. The oxidation facies may also be roughly recognized through current structures as well as through colour, composition and stratification of the sediment. In the absence of fossils the temperature can to some extent be estimated from the type of deposit (evaporites, special kinds of limestone), otherwise from reefs or glacial features. Coal beds, particularly in the case of a few seams which quickly wedge out, and certain tectonically produced salt deposits, are not suitable as temperature standards. An indication of the hydrofacies of a depositional area can also be obtained from certain bedding forms and through the presence of cyclic sedimentation.

Clay minerals are also suitable for direct standardization of the temperature facies if they have not suffered marked diagenetic alteration; kaolinite is often suitable in this respect. For the standardization of the supersaline facies talc, amongst others, is often used. The origin of the material can be deduced from the heavy minerals of the coarse fraction or from reworked fossils. The degree of diagenesis, in particular in the transition area to metamorphism, can also be read from the clay minerals. There is thus a range of possibilities for relating the geochemical data of facies analyses to known facies indices. Since the facies conditions are rarely constant over great distances and vertically may also alter quickly, several standardizations of the results of the analyses should be undertaken. In all cases great care must be exercised in making such standard comparisons since on them the accuracy of a geo-chemical facies investigation stands or falls.

GEOLOGICAL EVALUATION OF GEOCHEMICAL FACIES ANALYSES

With geochemical facies analysis in fossil-free sediments one can prove either a great deal or nothing at all since the most important control is absent. Investigations should therefore be started in fossil bearing zones and then later, with knowledge of the relationship of the minerals and fossils, be carried over into fossil-free sediment series. Investigation of single samples has therefore no value. In contrast it is important to determine the tendencies shown in the distribution of the investigated element with varying geological age. The geological questions are formulated in such cases as follows: what effect have (1) transgressions and regressions, (2) climatic development, (3) changes in the supply of sediment, (4) tectonic disturbances, and (5) diagenetic influences, on the distribution of chemical components in a single depositional zone? If these questions can be satisfactorily answered then one can proceed to detailed evaluation and decide from the available comparative standards to what extent and which facies may be differentiated bearing in mind the effects of transgressions, regressions, climate etc. through time. Thus most investigations, including those of the author, have so far been unsatisfactory: the hydrofacies of a large area or of a thick sequence has been decided from too few investigation points. Extension of the work has led to revision of previous studies.

An illustration of the care necessary is available from the paralic

Late Carboniferous of northwest Germany. Boron investigations were made there accompanied by applied geological work. The standardization of the boron values was made using fossiliferous material from Westphalian A and B. Through new outcrops created by the oil industry progressively younger Carboniferous beds were exposed and investigated. The division into hydrofacies then followed using the same boron values as in the older Late Carboniferous, but after many analyses it became clear that this standardization of the younger beds was erroneous, since the boron content rose markedly with increasing geological age. In this case identical boron content indicated quite different salinity divisions of the hydrofacies: in Westphalian A and B 0.040 % B_2O_3 is indicative of

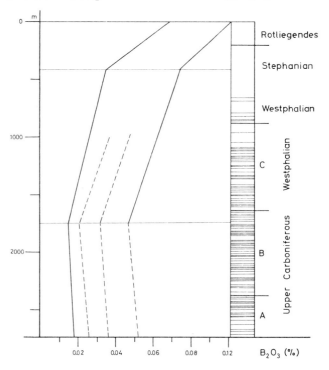

Fig.12. General increase in the boron content through the Late Carboniferous and Rotliegendes (Early Permian) of northwestern Germany.

a marine facies, whilst the same value in the Early Stephanian is found in limnic beds (Fig.12). The gradual transition to a hot, arid climate was probably responsible by causing increasing incorporation of boron in the sediments of evaporating remnant seas.

A further example is afforded by the effect of a small fault on the distribution of certain chemical components in the beds of the early fresh water molasse. In this case the analysis values are repeated.

The trends of the distribution of the elements can be presented in the following manner. In Fig.10 the most extreme low and high contents are limited by lines. These lines run with few exceptions parallel to one another through a considerable thickness of beds, and in some divisions change their orientation to the ordinate of the diagram. The separation of the bounding lines corresponds to the scatter of the values for an element in a given stratigraphical unit. The smaller the separation the more homogeneous the sediment may be (e.g., no marked fluctuation of the quartz or carbonate fraction). The parallel course of the maximum and minimum lines probably speaks for an equal increase or decrease of the investigated element in both the sandy and the more clay-rich intercalations. This can also be recognized from the parallelism of the boron content of the total sample and the fraction smaller than 2 μ; they differ only in magnitude.

An opposition of certain chemical combinations, which is shown by contrary inclinations of the bounding lines of different components (e.g., between B_2O_3 and $CaCO_3$) points to dependence on petrography and not hydrofacies. If one connects the maximal values with curves and not lines a further differentiation results (finely dashed curve in Fig.9, 10). Tangents can be drawn to the lowest points of these curves and the majority runs parallel to the line of minimum content. The tangents indicate approximately the median concentration of the element in the sediment. The boron curves lying above this line indicate the particularly high concentrations caused either by a special sediment make-up or by influence of the hydrofacies.

Thus from the course of such curves we can infer that directed trends are present connected either with the origin of the sediment, the quantity and rate of deposition of the sediment, or the hydrofacies. How may the influence of these different factors be recognized? An identical sediment source often affects a notable thickness of beds and widely distributed outcrops within a palaeogeographically limited depositional area. The lines and curves of the element distribution are relatively constant in such cases providing, of course, that the sedimentation rate experienced no marked change. An alteration in the hydrofacies may be indicated by various elements. In the case of the salinity facies in cyclic sediments alternating curves are obtained within a small thickness of beds; if connected with a marine transgression they may be correlated over many outcrops. V and Cr, like Fe^{2+}/Fe^{3+}, indicate changes of the oxidation facies, but since the aeration may radically change within a small area correlation of these curves from outcrop to outcrop is often not possible.

For Fig.9 an example of the relationship between the boron content and the subsaline facies of the Middle Keuper of Württemberg has been selected. There the boron content of the clay-marlstone is greatest where the thickest—now partly leached—beds of gypsum occur. Where the intercalations of gypsum cease at the bottom and top of the profile the boron content also drops sharply. This trend is also easily recognizable in the inclination of the concentration lines. Whereas the precipitation of gypsum indicates the highest degree of salinity of the south German Keuper Basin the Keuper River courses (Schilfsandstein, Kieselsandstein, Stubensandstein) stand out from the saline sediments through their lower boron content. This tendency is clearly seen in Fig.13. In each case the investigated sediment was a soft clay-marl, which occurs in the Kieselsandstein together with sandstone layers and pebbles. The same rules for evaluation can by no means be applied to all beds or elements. The examples from the Carboniferous of northwest Germany and the Keuper of southern Germany are only to show a method of how the results may be built into the existing geological picture. The evaluation in this method follows from the general (statistical) relationships. Only after the influence of transgressions

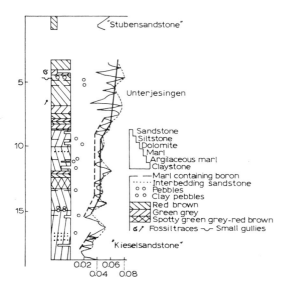

Fig.13. Increase in the boron content in the Middle Keuper from the Kiesel- to the Stubensandstein at Tübingen. (From ERNST, 1966; with kind permission of the editor.)

and regressions, climate, and diagenesis on the distribution of the elements are known should the resolution of local connections and the arrangement into facies of groups of beds be carried out. This requires a large number of analyses, but to attempt to do without them in deciding the hydro-, litho-, or biofacies of a broad depositional is to invite erroneous conclusions.

CHAPTER 5

Results of Geochemical Facies Analyses

The diagnosis of geochemical facies is dependent on numerous geological, petrographic, chemical and instrumental provisions (chapters 3 and 4). It requires permanence of the oceans and of the chemical components of the sea water, which can only be assumed from the end of the Precambrian and since then only for certain volatile (B, Br, Cl, F, J, SO_4) and organic components (amino acids amongst others). It requires also a degree of independence from climatic influences (weathering, evaporites) and thereby also from tectonic processes which can lead to the differentiation of vertical climate zones and evaporite basins. Al, Ca, K, Mg, (Na), (SO_4), Ti are especially affected and thus suffer during fluctuations in the earth's history.

Geochemical facies analysis is further dependent on the origin and rate of deposition of the sediment. Chemical elements already contained in considerable quantities in the source rocks can scarcely be overprinted by the hydrofacies of the environment of deposition. This applies to Al, K, (Na), Si, Ti, and also for the boron of tourmaline and recycled illites. Chemical elements strongly affected by diagenesis (through solution, reduction, exchange of bases, reconstruction of clay minerals) are wholly useless for geochemical facies analysis of older or deep lying sediments. Such elements are Ca, Fe, K, Mg, SO_4, amino acids (alteration through hydrolysis), and carbohydrates (bacterial decomposition).

Technical and analytical conditions have a not inconsiderable influence on geochemical facies analysis, in connection with the difficulty and the accuracy of the analysis. Amino acids are like most other organic components, more difficult to handle than, for instance, alkalis, earth alkalis, ferrides, halogens and heavy metals.

Of these inorganic elements several can be quickly analysed in serial processes (B, Br, Cl) whilst others (Al, Si, alkalis) often require a hydrofluoric acid disintegration if large modern apparatus is not available.

This review has already shown which chemical elements appear most useful for geochemical facies analysis and which are unsuitable. From the preceding sections it should also be clear where geochemical facies analysis is applicable and where not. (Applicable: clayey and organic sediments with few external facies characteristics. Inapplicable: sandstones; strongly metamorphosed or diagenetically affected sediments; sediments belonging to easily recognizable facies— used, however, in standardization.) The type of the sediment is also of importance, for instance boron determinations in kaolinitic material are not likely to yield significant information. In addition in the last section the errors consequent on too few analyses were dealt with. The most promising elements and possibilities for successful studies are therefore already very restricted. Consequently in the following sections only those chemical components which have either been sufficiently tested or may reasonably be expected to yield significant geochemical information with further experience. Studies based on an inadequate number of analyses and a too narrowly restricted area of investigation will only be mentioned for the sake of completeness at the end of the individual sections.

HYDROFACIES

Hydrofacies is understood as the effect of the physical and chemical character of the water on the sediment, and is applicable to marine and limnic conditions (chapter 2). It can be divided into: (*1*) salinity facies, (*2*) oxygen facies, and (*3*) temperature facies.

Salinity facies

Within the salinity facies of a sediment are gathered all the chemical characteristics produced by the salt content of the water in the area of deposition and preserved through the course of the development of the sediment. The majority of geochemical facies investigations up to the present have been in this field, one reason being the necessity of establishing facies divisions in fossil-free sediment comprising 25–30% of the beds (inclusive of evaporites) deposited since the end of the Precambrian.

Boron

The most investigated element in salinity determinations is boron. Since GOLDSCHMIDT and PETERS (1932) were able to recognize a difference in the boron content of marine and non-marine iron ores, interest has been directed more and more to this element which is present in sea water with a concentration of 4.8 mg/l (river water 0.01 mg/l). The first investigation attempting to distinguish marine from non-marine clay sediments was undertaken by LANDERGREN (1945, 1958) whose papers encouraged further facies investigations (BRADACS and ERNST, 1956; BRINKMANN and DEGENS, 1956; DEGENS et al., 1957; ERNST et al., 1958; FREDERICKSON and REYNOLDS, 1960).

A review made by HARDER (1963) shows the average boron content of various sediment types (Table II). In comparison with the content of various magmatic rocks listed in Table III these values are remarkably high (averages from HARDER, 1959). He also gave details of the boron content of minerals (Table IV).

The average values for clay sediments of various ages which Harder also compiled from figures given by a number of authors,

TABLE II

AVERAGE BORON CONTENT OF VARIOUS SEDIMENT TYPES

Rock	Range of variation B (g/t)	Average value B(g/t)
Clay and shales	25 – 800	100
Sand and sandstone	5 – 70	35
Limestone and marl	2 – 95	27
Dolomite	10 – 70	28
Iron ore	20 – 200	
Glauconite	350 – 2,000	
Possible average		85

TABLE III

BORON CONTENT OF VARIOUS MAGMATIC ROCKS

Rock	$B(g/t)$ *
Phonolite	5
Syenite, nepheline syenite	9
Granite and granodiorite	10
Diorite	14
Quartzporphyry, liparite	30
Andesite	20
Basalt, gabbro	2.5–6

* Conversion factor: 1 p.p.m. $B_2O_3 = 10^{-4}\% B_2O_3$
= 0.31 p.p.m. or g/t B.

TABLE IV

BORON CONTENT OF MINERALS

Mineral	$B(g/t)$
Paragonite	50 – 250
Muscovite	10 – 500
Serictic mica	2,000
Illite	100 – 2,000*
Montmorillonite	5 – 40
Kaolinite	10 – 30
Chlorite	max 50
Biotite	1 – 6
Quartz	0 – 35

Olivine, garnet, zirkon, topaz, feldspar lie in part beneath the boundary of detection

* Fine grained illite contains more boron.

are no longer representative. Corresponding recent analytical results are given in Table V.

Metamorphic processes, both contact and regional, reduce the boron content (ERNST et al., 1958; HARDER, 1959; ERNST and LODEMANN, 1965).

Boron reaches the sea as a constituent of weathering solutions and volcanic exhalations. The high incorporation quota of this element in mica minerals, especially illite and glauconite, and furthermore a relatively high adsorption on exchange-susceptible clay minerals maintains a rough balance between introduction and

TABLE V

CONTENTS OF BORON IN SEDIMENTS OF DIFFERENT FACIES

Deposit	B_2O_3 (wt. %)	Location
Quaternary	0.010 – 0.030	Baltic Sea, northern Germany
Tertiary	0.020 – 0.040	lower and upper Rhine, northern foreland of the Alps
Cretaceous	0.030 – 0.040	northwestern Germany
Triassic		
Keuper	0.030 – 0.050	southwestern Germany, Luxembourg
Muschelkalk	0.030 – 0.040	southwestern Germany, Luxembourg
Buntsandstein	0.030 – 0.050	Germany
Upper Permian	0.040 – 0.060	northwestern Germany
Lower Permian	0.040 – 0.080	northwestern Germany
Upper Carboniferous	0.015 – 0.050	Ruhr and Saar district (Germany), Belgium, Austria
Lower Carboniferous	0.015 – 0.040	Germany
Devonian	0.015 – 0.040	Eifel (Germany)

fixation of boron (HARDER, 1961; REYNOLDS, 1965a, b).

HARDER (1961) was able to prove the incorporation in mica and illite through experiment. This incorporation is temperature dependent and requires a relatively long time before saturation occurs. Harder assumed that Al in the easily accessible outer tetrahedral positions was substituted by boron. In contrast WALKER and PRICE (1963) believed the substitution of Si by B in illite and muscovite to be possible. The linkage of boron to illite was also accepted by FREDERICKSON and REYNOLDS (1960) and in part by STADLER (1963). Because of this close linkage various authors have related the boron content to illite (PORRENGA, 1963; WALKER, 1964; SINGH, 1966; HELING, 1967a) or to the K_2O content of the illite (WALKER and PRICE, 1963; CURTIS, 1964; REYNOLDS, 1965a, b), see Fig.14. Through the firm incorporation of the boron in mica it is relatively resistant to weathering. It can scarcely be won from the clay fraction with concentrated hydrochloric or nitric acid, but is accessible to strong alkali solutions (DEGENS et al., 1957; GOLDBERG and ARRHENIUS, 1958).

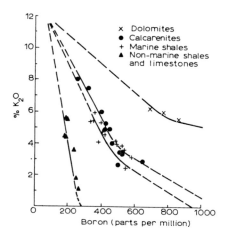

Fig.14. The B/K_2O ratio in various rocks. (From WALKER and PRICE, 1963, p.835, fig.1; with kind permission of the editor.)

The whole boron content of the sediment cannot, however, be related to the mica. Grain size separations of clays show that two maxima of boron content occur in this class (HARDER, 1959): one

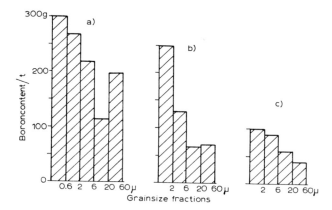

Fig.15. Boron content of various grain-size fractions of marine clays: (a) Rhaetic clay near Göttingen (Germany); (b) Liassic (Planorbis bed) near Göttingen; (c) recent deposit in the mouth of the Amazone (Meteor Expedition). (From HARDER, 1963, p.244, fig.1; with kind permission of the author and editor.)

lies in the clay fraction proper below 2 μ, the other, with a smaller proportion, in the region of 20 μ (Fig.15, 16). A part of the boron in

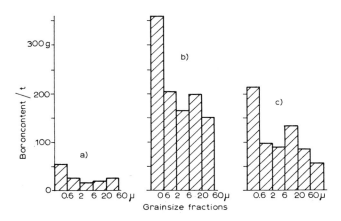

Fig.16. Boron content of various grain-size fractions of limnic clays: (*a*) banded clay, Värmland (Sweden); (*b*, *c*) Pleistocene clay (glacial) near Bilshausen (Germany). (From HARDER, 1963, p.244, fig.2; with kind permission of the author and editor.)

the coarser fraction is certainly traceable to tourmaline. A further part is in the opinion of LEVINSON and LUDWICK (1966) also bound adsorbed on clay minerals, and is dependent on the nature of the adsorbing clay mineral: considerably higher boron contents are known from montmorillonitic tuffs than from kaolinite (WINTER, 1969).

Boron is also bound to plants and to animal hard parts. Coals are generally poor in this element but in plant ash the content is higher (JANDA and SCHROLL, 1959; KEAR and ROSS, 1961). The concentration in plants rises according to SWAINE (1962) in the transition from the limnic to the marine regime (marine influenced coals). The incorporation of boron in monomineralic shells (e.g., *Cardium*) is also dependent on salinity (LEUTWEIN, 1963).

According to EAGAR (1962) the concentrations of organic substances and boron run antipathetically. In contrast complex boron (NOAKES and HOOD, 1961) varies sympathetically with the oxygen content of the water. In the cases of an O_2 minimum the correlation between complex boron and salinity is disturbed.

The dependence of the content of boron on the source area of the sediment has been demonstrated for the Schilfsandstein of the German Keuper by HELING (1967) and for the English Late Carboniferous by SPEARS (1965). Spears traced the varying B/K_2O ratio to differing weathering effects. However, HIRST (1962b) considered that the differing boron content might be connected with a fluctuating rate of sedimentation.

In NICHOLLS's (1963) opinion the use of boron for salinity analysis is supported not only by chemical and physical hypotheses but also by statistical information. LEVINSON and LUDWICK (1966) came to similar conclusions and accepted the boron content of unfractionated clay-stone could be used as a salinity index. The statistical results of ERNST (1964) for the Carboniferous of the Ruhr, BAUMANN (1968) for the Tertiary sediments of the unfaulted molasse of southern Germany and GITTINGER (1968) for the boundary area of the Late Muschelkalk–Early Keuper in Luxembourg, may help to clarify the problem of the facies dependence of boron (investigations carried out by the Tübingen Geological Institute). Fig.17 illustrates the distribution of boron in relation to the coal-bearing terrestrial facies and the marine ingressions in the northwest German Carboniferous. The lower boron values are restricted to the coal-yielding zones, whilst with the approach of marine horizons (Katharina, Westphalian A/B, Ägir, Westphalian B/C) they rise. A general increase in the boron content from Westphalian C beds onward is superimposed on the salinity-dependent boron (Fig.12 on p.60). However, the difference between the various salinity zones is still recognizable at least as far as the early part of Westphalian D, as the investigation of fossiliferous sediments in the boundary area between Westphalian C and D near Ibbenbüren (Westphalia) has shown. This tendency holds good for the lower Late Carboniferous in other areas.

General trends in the boron content in connection with the original distribution of land and sea can also be obtained from a statistical evaluation of the content of boron and other chemical

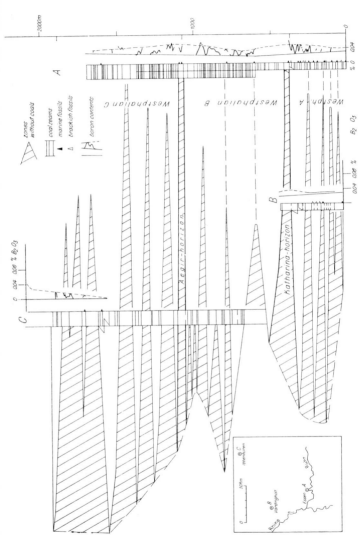

Fig.17. Distribution of the boron content in terrestrial and marine facies of the northwest German Upper Carboniferous (= calculation from several thousand analyses).

components in the unfaulted molasse of southern Germany. In the Muschelkalk and Keuper of Luxembourg there is a statistical relationship between boron and the basin and marginal facies.

As already mentioned such salinity-dependent trends can be overprinted by the effects of the temperature facies. Fig.18 shows

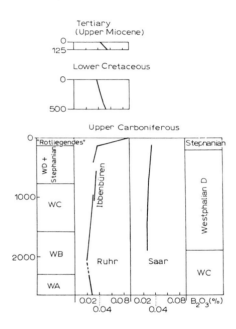

Fig.18. Distribution of the boron content in various systems in Germany. (From ERNST, 1966, p.25, fig.1; with kind permission of the editor.)

two relevant examples from the Late Carboniferous of the Ruhr and Saar in which the minimum boron content rises markedly with the change to a hot, arid climate. In the Early Cretaceous of Emsland (Lower Saxony), on the other hand, the boron content decreases from the latest part of the Jurassic to the Valanginian, which is apparently in accord with measurements of palaeotemperature (ENGST, 1961). In the Late Miocene of the upper Rhine Graben (Germany) the boron values sink with the approach of the cooler Pliocene. Each of these investigations is based on several hundred analyses of the relevant formations.

How far diagenesis can overprint the boron values in the same sense as the effect of temperature has not yet been unequivocally established. Many samples have been taken from the Carboniferous in borings at depths of up to 4,000 m and their boron values are not significantly different from others taken from the same beds where tectonism has not placed them at such depths. Changes in sedimentation, for instance through a gradual change in the source area, are out of the question for the Early Cretaceous beds of Emsland (eastern Lower Saxony) as well as for the Late Tertiary beds of the upper Rhine Graben. However, such a change cannot be wholly ruled out for the Late Carboniferous of northwest Germany, since material from both Carboniferous beds folded in Westphalian C and Ordovician from the crystalline periphery of the North German Basin was carried into the Late Carboniferous trough. A change in the source of the sediment is manifested as a very small thickness of the Saar Carboniferous, where the Holz Conglomerate is characterized by an accentuated boron maximum (Fig.19). With the Holz Conglomerate, sediment begins to be supplied from the north in the Saar Carboniferous (KNEUPER, 1964) and boron-rich material was probably brought into the area of deposition. Connection with a salinity change cannot be accepted since limnic conditions were prevalent in the Late Carboniferous of this region.

Statistical methods have thus shown that the influence of salinity or the derivation of the sediment can clearly stand out from the general background of the boron distribution. However, if, from other geological data, one is unable to say in which salinity milieu the deposition of a thick sedimentary series took place, then a temporary change in the origin of the sediment (e.g., the Holz Conglomerate) if judged only by the boron content may be mistaken for a marine horizon.

The earlier optimism over the suitability of boron as a salinity indicator can therefore no longer be fully supported. Boron is only an index of salinity in certain special areas of deposition charac-

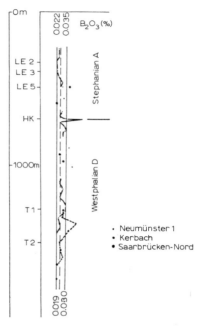

Fig.19. Boron content of reworked, non-carbonaceous sediments in the Holzer conglomerate (Late Carboniferous, Westphalian D—Saar region, Germany.)

terized by regular subsidence, a uniform rate of sedimentation, and a single source of sediment. This applies to the Subvariscan fore-deep for the period between the latest Namurian and the end of Westphalian B. In such cases the influence of temperature is not able to over-print the facies dependence of the boron.

The younger Alpine fore-deeps of Tertiary age contain, in contrast, sediments of varying derivation, traceable to the crystalline core and even individual Alpine nappes. Moreover they show in part a high rate of sedimentation and often reworking and erosion of the bottom sediment by marine currents or rivers. In these areas determination of the salinity with the aid of boron is burdened with uncertainty.

Further papers not previously cited which deal with certain special aspects the occurrence of boron, or which are only restricted investigations are MOTOJIMA et al. (1960), SHERBAKOV (1961), TOURTELOT et al. (1961), SPJELDNAES (1962), STAVROV and KHITROV (1962), BUGRY et al. (1964), and ISHIZUKA (1965).

Bromine

The bromine content of sediments has not been sufficiently investigated. Sea water contains about 65 p.p.m., and the Cl/Br ratio is about 300. In the bodies of animals the ratio according to KREJCI-GRAF (1963c) is fairly constant between about 500 and 1,000. In plants, in particular those of marine origin, Br may be considerably enriched (in *Salicornia herbacea* some 21,700 p.p.m. as opposed to 262 p.p.m. in dried fresh-water plants). KREJCI-GRAF and LEIPERT (1936) gave further details of the Br content of some sediments (Table VI).

TABLE VI

BROMINE CONTENT OF SOME SEDIMENTS

Sediment	Br (p.p.m.)
Peat	up to 30
Bituminous clay	up to 33
Interglacial clay (Helgoland)	33
Late Pliocene oil-shales (Valta Fopor, Romania)	29
Marine algal gyttja	132
Zechstein dolomite (Volkenrode, Thuringia)	16
Triassic Hauptdolomit (Seefeld, Tirol)	1.77

The geochemical importance of Br lies on the one hand in tracing the origin of formation waters, and on the other in investigations of evaporites. Bromine is strongly enriched in residual liquors and can be firmly incorporated in sylvine, carnallite, and bischofite in place of Cl, in concentrations corresponding to its enrichment in the liquor at the time of formation of the minerals (BOEKE and EITEL, 1923; D'ANS and KÜHN, 1944), the last formed containing the greatest quantity of Br. This element is therefore not only an index of supersalinity but also of the relative stratigraphic position of the salt (Fig.20).

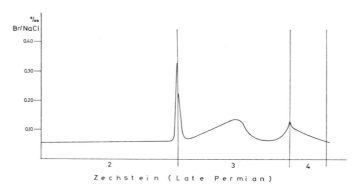

Fig.20. Distribution of the bromine content in the central German Zechstein (Late Permian) in relation to stratigraphic position. (After SCHULZE, 1960; with kind permission of the editor.)

Chlorine

Chlorine is a major constituent of sea water, and amounts to 19,000 p.p.m. in comparison with 8.3 p.p.m. in river water. It is not so strongly enriched in organisms as Br or I, but its concentration in lower marine organisms is greater than in some fresh-water species (RANKAMA and SAHAMA, 1950). During weathering Cl passes easily into solution. A further part is fed into the oceans through exhalation. A not inconsiderable quantity of Cl is combined with K and Na to form salt beds which have fixed about 5% of the present NaCl content of the oceans.

According to the early investigations made by KURODA and SANDELL (1950), BEHNE (1953), and CORRENS (1956) the water insoluble Cl in sediments is present in a primary combination as the hydrosilicate, which JOHNS (1963) found to be linked to chlorite minerals. This process occurs during the transformation of montmorillonite in the presence of high salt concentration. GRIM and JOHNS (1954) observed this transformation of montmorillonite into chlorite and illite with rising pH and salinity near Rockport on the coast of Texas. The river Guadalupe builds there a delta into the protected bay of San Antonio, which is in turn connected with the Gulf of Mexico. JOHNS (1963) recorded the chloritization as occurring through the saturation of Ca montmorillonite in Mg by cation exchange in sea water. In this process brucite islands are formed step by step between the silicate layers (montmorillonite-chlorite mixed layer mineral), which gradually grow together until a chlorite mineral has been created. During chloritization Cl and OH ions are incorporated together in the intermediate layers, and the OH ions are later gradually driven out by Cl ions.

In contrast to the chlorite minerals of fresh-water sediments with less than 100 g Cl/t, the authigenic chlorites of the marine facies contain up to 1,000 g Cl/t. The increase of the Cl content with the transformation of montmorillonite with rising salinity is shown in Fig.21. The investigation is based on material of a grain size less

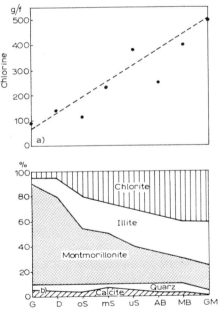

Fig.21. Chlorine distribution (*a*) and mineral composition (*b*) of the fraction $<1\mu$ in sediments of the Guadalupe River and associated bays in the Gulf of Mexico near Rockport (U.S.A.). Increasing salt content from left to right. (From JOHNS, 1963, p.225, fig.9; with kind permission of the author and editor.)

than 1 μ. The analyses of the coarser fraction of the sediment showed no connection with the salinity of the milieu. No information was obtained which revealed how the detrital chlorite brought by rivers behaved on meeting sea water. Similarly the question of what happens to the linked chlorine if a recent sediment undergoes diagenesis was left unanswered. JOHNS (1963) mentioned the possibility that the release of Cl, perhaps together with F, may be the starting point for hydrothermal solutions. In the meantime, however, it has been possible to show that the same proportion of chlorine is present in marine and non-marine Cretaceous beds in the U.S.A., which speaks for a certain stability of the Cl in the chlorites (W. D. Johns, personal communication, 1968). This result accords with the supposition already mentioned in the chapter on diagenesis, in which it is stated that not only is Cl plentifully available to the sediment from the sea water, but that as far as the shallow burial stage, it is not used in diagenesis. This fact suggests that Cl may be a good indicator of palaeosalinity. Statistical evidence for this presumption from other areas is, however, not yet available.

Alkalis and earth alkalis

In the section "Permanence of the Oceans" it was mentioned that in the third phase in the development of the oceans more Ca than Na passed into solution and was supplied to the oceans. Moreover Na forms compounds with Cl^- and SO_4^{2-} ions, which led during the major salt forming periods of the earth's history to the consumption of about 5% of the reserves in the oceans. Na is therefore from the outset an unsuitable element for geochemical facies determinations since the prerequisite permanence of element concentration in the oceans is not satisfied.

Na enrichment in marine clays was reported by MICHALIČEK (1961), but related information for the limnic milieu is not available. The Na cycle is represented in Fig.22. A part of the Na content of the oceans is lost through adsorption and in the pore solution of sediments. After emergence, folding and weathering of such sediment a part of this Na is returned once more to the ocean via rivers. A smaller part, however, remains on the continent as salt crusts,

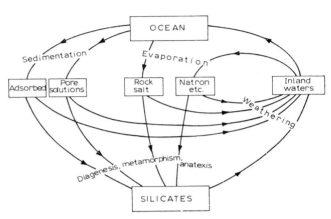

Fig.22. The sodium cycle. (From GREGOR, 1967, p.48, fig.13; with kind permission of the author and editor.)

but considerably more Na is used up in oceanic salt deposits.

Through diagenesis and metamorphism notable quantities of salt from all three parts of the cycle (sediment, continental and oceanic salt) become involved in silicate formation. In agreement with WEDEPOHL (1963) this pattern of distribution results in a gradual decrease in transport of Na to the oceans. In contrast CARSTENS (1949) assumed that the rate of removal of sodium salt from the oceans was similar in the past to that of the present day (Fig.23).

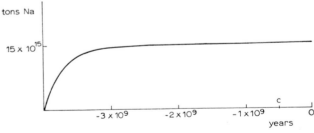

Fig.23. Sodium content of the ocean through geological time (C = Cambrian) according to CARSTENS (1949). (From GREGOR, 1967, p.33, fig.8; with kind permission of the author and editor.)

The Na content may fluctuate strongly over short periods of time, as shown in investigations made by CAMERON (1966) in Early Cretaceous beds in Alberta (Canada). Thus beds containing volcanic detritus contain 2.37%, and the lower and upper parts of the Early Cretaceous 0.21% and 0.87% respectively. The value for the sediment with volcanic detritus agrees closely with the figure of 2.83% given for RANKAMA and SAHAMA (1950) for igneous rocks. In the Gulf of Paria (between Venezuela and Trinidad) the average content of Na in the various sediments is as follows: recent delta sands, 0.43; platform sands, 0.88; green mud, 2.05; blue mud, 1.48; delta clay, 1.08%.

The relatively high Na content of these clays can probably be related here also to sediment supply from a hinterland of igneous rocks, in this case the Guyana Shield in which high grade metamorphics are also present. Dependence on salinity can be ruled out in this example but in contrast the incorporation of Na in recent biogenic carbonate proceeds according to the salinity. A high Na content in fossil aragonite shells (Na more than 0.3%) indicates a marine milieu and a very low concentration (below 0.05%) fresh water (G. Müller, personal communication, 1968).

According to VINOGRADOV and RONOV (1956) in the sediments of the Russian Platform K decreases constantly from the Proterozoic to the Quaternary. Like Na periodic incorporation in salt beds also removes quantities of K from the oceans. In addition during diagenesis K is strongly adsorbed on clay minerals, and is also used in the reconstruction of micas. The K content of igneous rocks at 2.59% is somewhat less than that of Na (RANKAMA and SAHAMA, 1950). On average, however, the proportion of K in the recent sediments of the Gulf of Paria is higher than Na. Plants are enriched in K and are burned to provide potash. From these examples potassium cannot be expected to have great importance for the analysis of salinity facies.

Local Li enrichment in NH_4NO_3 extractions was reported by KREJCI-GRAF and ROMEIS (1962) from limnic lime-free as opposed to lime-poor clays. The reason is probably that Li is more firmly bound in marine sediments.

The distribution of alkalis and earth alkalis in sediments of varying salinity grades is given in Table VII.

TABLE VII

DISTRIBUTION OF ALKALI AND EARTH ALKALI ELEMENTS IN
SEDIMENTS OF DIFFERENT SALT-CONTENT[1]
(Average contents, wt. %)

Recent sediment	Na	K	Ca	Mg	Location
Delta sands	0.44	0.93	0.46	0.32	inner delta of Orinoco
	0.50	0.79	0.26	0.30	
	0.35	0.50	0.20	0.12	
Platform sands	0.82	0.70	13.22	0.44	S Trinidad
	0.81	0.87	6.53	0.54	
	0.74	0.78	6.43	0.55	W Trinidad
	1.26	1.29	1.13	0.69	
	0.89	1.30	0.80	0.40	
	0.96	1.13	0.84	0.65	
	0.73	0.91	0.43	0.34	
	0.83	0.93	0.64	0.41	
	1.11	1.01	1.11	0.39	
	0.41	0.51	0.75	0.90	NW Trinidad
	1.15	1.09	0.86	0.63	
	0.83	0.86	0.85	0.48	
	0.76	1.53	9.52	1.01	
	2.56	1.85	0.58	1.66	
	1.83	2.01	0.70	1.54	
Clays	1.08	1.64	0.51	0.69	delta of Orinoco
	2.37	1.99	0.79	1.30	W Trinidad (20–60 cm below surface)
	1.62	1.95	1.11	1.78	W Trinidad
	2.43	1.89	0.56	1.41	central part of Gulf of Paria

Tertiary		Na	K	Ca	Mg	Salinity of deposit
Upper Pannonian		0.62	1.98	5.18	0.77	< 3
Middle Lower Pannonian		0.80	1.69	2.96	1.02	5 – 12
Lower Pannonian		0.56	1.28	4.99	1.49	5 – 10
Sarmatian horizon	1 – 4	0.62	1.51	6.37	0.88	15
Sarmatian horizon	5 – 8	0.41	1.38	8.19	1.42	15 – 20
Sarmatian horizon	9 – 13	0.71	1.86	3.20	0.59	20 – 25
Sarmatian horizon	14 – 17	0.64	1.31	7.53	1.50	20 – 25
Sarmatian horizon	19 – 20	0.64	1.27	13.50	4.21	25
Upper Tortonian	1 – 3	0.59	1.56	4.95	0.84	25 – 30
Upper Tortonian	4 – 11	0.82	1.52	4.64	0.67	30 – 35
Helvetian		0.54	1.83	5.98	0.24	35

[1] First part from HIRST (1962), p.317); second part: Austrian samples—Vienna
Basin—from KREJCI-GRAF et al. (1966, p.73, 88).

The Ca/Mg may be usable for judging the original salinity (Ca documents terrestrial and Mg marine origin), if suitable possibilities for absorption by the sediment were prevalent. In peats WERNER (1963) recorded Ca and Mg to be clearly enriched in comparison with Na and K. A Ca/Mg ratio of approximately 15 in brown coals indicates limnic influence and below 5 a marine influence.

However, it must be mentioned that the Mg/Ca ratio is apparently dependent on age (VINOGRADOV, 1957). According to CHILINGAR (1953) and DALY (1907) the average Ca/Mg ratio in the U.S.A. and Canada decreases with age.

The Ca/Sr ratio is a valid index of salinity. According to KREJCI-GRAF and ROMEIS (1962) it is higher in clays and marls of marine deposits than in limnic sediments. LANDERGREN and MANHEIM (1963) were also of this opinion for present day deep sea clays and the sediments of the Kattegat between Denmark and Sweden, and in the Baltic. On the other hand TUREKIAN (1964) established only a decrease in the Ca/Sr ratio with time, and ODUM (1950) confirmed a fall in Sr content in beds from the Ordovician to the Late Carboniferous. TUREKIAN and KULP (1956) summarized the factors on which Sr is dependent as follows:

(*1*) Sr is dependent on the Cr/Ca ratio of the liquid phase.

(*2*) The presence of calcite or aragonite is decisive for the Sr content.

(*3*) The vital activity of organisms influences the Sr content.

(*4*) The temperature and salinity of the sea water plays a role in the distribution of Sr.

(*5*) Diagenesis may alter the contained proportion of Sr.

The effect of aging on the Sr/Ca ratio was thought by KREJCI-GRAF (1966) to be a local effect due to differing sediment source. STADNIKOFF (1958) used the ratio Na^+/Ca^{2+} ions in order to determine salinity. According to his investigations marine clay stones contain dominantly Na and only subordinate Ca as exchange susceptible cations, whereas under limnic conditions the roles are reversed and the Ca^{2+} ion is dominant, and Na is scarcely present. The ratio of the amount of exchange susceptible Na^+ and Ca^{2+} ions (in milli-equivalents) can be determined through the treatment of the clay stone with a special solution of barium chloride. The resulting salinity coefficients, or ratios between ionic Na and Ca, are as follows: fresh-water beds 0.2–2.9; for clay with *Anthraconauta* 0.7–4.8; for clay with *Lingula* 1.8–4.9; and for marine clay stones 0.5–10.6. However, the standardization of the analyses is not exact, and the overlap of the values is so great that in practice it is impossible to decide which facies is present (a value between 1.8 and 2.9 still leaves a choice of facies ranging between fresh and marine). Furthermore, attempts at verification of these figures by KREJCI-GRAF et al. (1965) and by ERNST and WERNER (1960) using the same procedures were negative.

Strontium and barium are enriched during weathering. In shales the amount of Sr is 100 times, and Ba 1,000 times that of sea water (BARTH et al., 1939). Limestones preferentially concentrate Sr (Sr 425–850 p.p.m., Ba 270 p.p.m.), whereas clays concentrate Ba.

Of the alkalis and earth alkalis only the Ca/Sr ratio is suitable for the analysis of salinity, but it has not yet been sufficiently statistically investigated. The Ca/Mg ratio only plays a role in the case of organic material, which is notable for a high propensity for adsorption (peat). Moreover when dealing with alkalis and earth alkalis the method of chemical extraction (methanol; NH_4NO_3) is also decisive with respect to which cations go preferentially into solution. From a methanol extraction one can use the Na/Cl ratio to separate either limnic from kaspi-brackish samples or normal brackish from marine (KREJCI-GRAF et al., 1965).

Sulphur and isotopes of sulphur, carbon and oxygen

Sulphur as SO_4, like boron, bromine, and iodine, is a volatile element the quantity of which in the oceans cannot be accounted solely to weathering. In the case of sulphur RICKE (1963) concluded that only 5% could have been derived from the weathering of eruptive rocks. The remainder can only be traced to volcanic exhalations and hot springs. This independence of sulphur from weathering and its ability to form a wide variety of compounds because of its differing oxidation stages, make this element of particular interest for the geochemist. According to RICKE (1963) the average percentages of sulphur in the rocks of the crust are: eruptives, 0.03; sediments, 0.20; sediments inclusive of evaporites, 0.40; sea water, 0.09; fresh and inland water, 0.0006%. The S content of inland waters is in this survey about two powers of ten less than in sea water. Nevertheless, present-day river water contains considerably more S than can have resulted from weathering. Ricke traced this excess to human activity (fertilization, burning of coal and oil), saliferous spray from the oceans, and its introduction by hot springs and volcanic exhalations. How is this contrasting distribution of sulphur manifested in the sediments? Details of the distribution of sulphur in marine and limnic clay stones are given in Table VIII. In this review the amount of sulphur

TABLE VIII

DETAILS OF THE DISTRIBUTION OF SULPHUR IN MARINE AND LIMNIC
CLAY STONES (After RICKE, 1963, table I)

Clay stone	S (%)	Reference
15 marine shales (Late Carboniferous)	0.92 ± 0.68	RICKE (1963)
16 clayey shales (Late Carboniferous; Namurian C, main seam)	1.29 ± 1.50	RICKE (1963)
17 marine shales (Late Cretaceous; Pierre shale)	0.55	TOURTELOT (1964)
15 fresh water shales (Late Carboniferous)	0.15 ± 0.13	RICKE (1963)
9 fresh water shales (Late Carboniferous; Westphalian A)	0.19 ± 0.21	RICKE (1963)

in marine and non-marine shales is different. If one in addition compares the detailed distribution of this element in selected horizons of the same depositional area (Table IX) the differentiation outlined below can be recognized.

In the marine horizons above the main seam in the mine "Rudolph" of the Ruhr studied by RICKE (1963), only the lower 1.75 m are characterized by a high content of sulphur (more than 3%), although judging by the fauna *Planolites ophthalmoides* the salinity of the original environment remained unchanged through beds up to 2.75 m above the seam. In beds above 1.75 m, where in

TABLE IX

SULPHUR CONTENT (%) OF CARBONIFEROUS SHALES
(From RICKE, 1963, p.276)

Sample	Shale-character	Total S	Sulphate-S	Sulphide-S
Limnic sediments				
Westphalian A	bituminous	0.14	—	—
(Bochumer) Schich-	shale with coal	0.69	—	− 11.5
ten), mine "M.		0.08	—	—
Stinnes ¹/₂",	sandy	0.27	—	− 11.4
0 – 1.20 m around	sandy	0.32	0.02	− 11.0
coal-seam "Albert"	sandy	0.07	—	—
	sandy	0.07	0.01	− 11.9
		0.08	0.01	− 13.0
		0.05	—	—
Marine sediments	bituminous	3.0	0.06	− 2.9
Namurian C	bit.	1.4	—	—
(Sprockhöveler	bit.	3.2	0.39	− 34.9
Schicht), mine	bit.	2.9	—	—
"Rudolph", 0 – 4 m	bit.	3.0	0.38	− 35.9
around coal-seam	shale with coal	3.6	—	—
"Hauptflöz"	bit.	3.3	0.40	− 35.2
		0.05	—	—
		0.06	—	—
		0.03	—	—
		0.03	—	—
		0.04	—	—
		0.04	—	—
		0.02	—	—
		0.04	—	—
		0.04	—	—
15 marine shales, Upper Pennsylvanian (Clarion Formation)		0.92 ± 0.68*		
15 limnic shales, Pennsylvanian (Freeport Formation)		0.15 ± 0.13*		

* After M.L. Keith and A.M. Bystrom in: RICKE (1963, p.273).

addition the burrows mentioned as characteristic of brackish water dominate, the amount of sulphur sinks to an average of 0.04%. This is precisely the average value for the limnic beds above the seam "Albert 1" in the Westphalian A (M. Stinnes 1/2 Mine, Ruhr, see also Fig.24).

In this example either the fauna departs from the salinity conditions associated with it in other beds in the sequence, or the S content has been affected by factors other than salinity. The oxygen facies may have exerted an influence since in still-water regions a lack of oxygen (low redox potential) may lead to a marked precipitation of iron and other heavy metal sulphides (see p. 97) in dark, bituminous sediment.

The profile used by Ricke for standardizing the S content comprises dark grey to black bituminous clay stones in which pyrite and marcasite are restricted to the first 1.5 m above the seam (see also ERNST, 1963). In this part of the profile the maxima for the S content lie above 3%. In the limnic beds above the Albert seam the lower 0.3 m also consist of dark grey bituminous clay stone which passes upwards into normal middle grey clay and siltstone. In its central part the bituminous layer has an S content of 0.69%, but in the succeeding less bituminous part only an average percentage of 0.07 is reached.

It is thus clear that the distribution of sulphur in the Late Carboniferous beds described by Ricke is in part determined by the original distribution of oxygen in the sediment. A similar case is known from the boring "Tettnang 1", north of Lake Constance (southern Germany), where BAUMANN (1968) reported double the percentage of S (up to 0.020%) in the organically influenced parts of the fresh-water molasse in comparison with an average of 0.010 in beds deposited under more oxygenated conditions. The V/Cr and Fe^{2+}/Fe^{3+} ratios rose in this profile together with S.

Therefore we must bear in mind in view of these examples that high sulphur content may be traceable to the oxygen facies. Similarly we must reckon with the re-solution of sulphides and sulphates on changes in pH during diagenesis, and consider that sulphate-bearing pore water may migrate and cause further diagenetic chan-

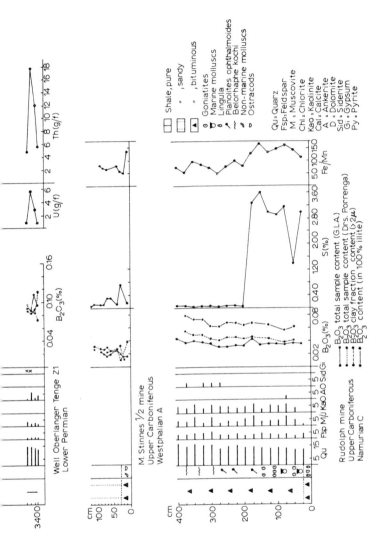

Fig.24. Relation between rock composition, boron content (clay fraction $<2\mu$ and total sample), sulphur, and the Fe/Mn ratio in marine beds (Rudolph Mine, Namurian C) and non-marine beds (M. Stinnes $^{1}/_{2}$ Mine, Westphalian A) of the Ruhr Carboniferous. (From ERNST, 1963, p.354, fig.1; with kind permission of the editor.)

ge and sulphur enrichment (RICKE, 1963). Both these factors reduce the merit of sulphur as an index of salinity, and detailed attention must be paid to the roles they might have played if it is to be used as such.

Systematic investigation of the ratio of the sulphur isotopes in marine and non-marine sediments has only been carried out on a small scale. The isotope exchange reaction leads to heavy sulphate and light sulphide. Present day sea water sulphate has a value of $\delta = +20.1 \pm 0.3$ according to THODE et al. (1961). Rain water has, in contrast, a delta value of -1.7 to $+6.1$ (THODE, 1949; OSTLUND, 1959) and river and lake water from $+2.8$ to $+11$ (NAKAI and JENSEN, 1960). According to studies made by Nakai and Jensen the difference in the ratio between the sulphur of the sulphate and sulphide (produced by reduction) in fresh-water sediments is less than in marine deposits. RICKE (1963) supposed, on the basis of laboratory experiments performed by HARRISON and THODE (1958a, b), that there is a smaller sulphate availability in fresh-water deposits, thus providing different living conditions for the bacteria. Similarly the concentration and type of organic substance and chloride content of the water affect the activity of the bacteria and the ratio of the isotopes they produce. Ricke expected a significantly lighter sulphur in fossil sediments than in present day sea water. According to present knowledge (HOLSER et al., 1963; NIELSEN and RICKE, 1964; THODE and MONSTER, 1964; NIELSEN, 1965) the delta value of ^{34}S of sea water was somewhat higher from the Cambrian to the Silurian than it is today; in the Permian it was at a minimum of 11% and rose again during the Mesozoic to its present level (MÜLLER et al., 1966).

In the pore water sulphate of older sediments the ratio of the amount of the S isotopes may be altered by bacterial sulphate reduction. MÜLLER et al. (1966), investigating such pore water, have shown that both within and between different geological systems (Permian, Jurassic, Cretaceous and Tertiary of northern and southern Germany) important differences exist. In several cases the S isotope ratio is affected by sulphate evaporites.

If we look at the few isotope ratios given by RICKE (1963) for the already mentioned Late Carboniferous sediments, we find values for δ ^{34}S sulphide of —34.9 to —35.2 in marine sediments in comparison with —11.0 to —13.0 in the limnic facies. Values for δ ^{34}S sulphate could only be calculated for the marine clay stone (—23.9 to —24.7) since no sulphur could be detected in the limnic sediments of the Late Carboniferous.

It seems, therefore, that the distinction of marine and non-marine sediments of the Late Carboniferous of the Ruhr may be possible using sulphur isotopes even if the sulphate itself gives no hint of being significantly divided between the various facies (see above). Unfortunately the broad ranging sampling necessary is scarcely possible because of the cost of the complicated laboratory procedure.

Having dealt with the isotopes of sulphur it is appropriate to discuss here the other isotopes which may be usable in salinity analysis: D, ^{18}O, ^{12}C, ^{13}C. The isotopes of nitrogen are fractionated only in the transformation of organic substances. B and P isotopes have not yet been sufficiently investigated in this connection.

Isotope ratios are in general influenced by the environmental temperature, evaporation and the nature of the melt which produced possible source emanations (CLAYTON, 1959; HORIBE and KOBAYAKAWA, 1960). ^{12}C is enriched in limnic in comparison with marine carbonates (SILVERMAN and EPSTEIN, 1958; CLAYTON and DEGENS, 1959; KEITH and DEGENS, 1959; WEBER, 1964a). Marine mollusc shells were found by KEITH et al. (1964) to have δ ^{13}C values between + 4.2 to — 1.7‰ in contrast to —0.6 to —15.2‰ for shells from a limnic environment. Fresh-water organisms contain less deuterium, ^{18}C and ^{18}O than sea water (RANKAMA, 1954). The ^{12}C/^{13}C ratio in coals and rocks bears no relation to the salinity facies (KREJCI-GRAF and WICKMAN, 1960). Further mention of the use of isotope analysis can be found in the sections on oxygen and temperature facies.

Phosphate

NELSON (1967) published a method for determining palaeosalinity with the aid of the ratio Ca phosphate / (Ca + Fe) phosphate. The method is founded on the distribution of the phosphates of Al, Fe, and Ca in soils and sediments. In marine sediments Nelson believed only Ca phosphate to be present whilst in soils and limnic deposits Al and Fe phosphates are the most important phosphor compounds.

Checking of this method by MÜLLER (1969) in Carboniferous, Permian and Tertiary sediments revealed that it is not universally applicable. The reasons lie on the one hand in erosion of outcropping marine strata and on the other in diagenesis. If fresh water sediments are brought during diagenesis under reducing conditions Müller found that Fe phosphate is reduced to FeS and phosphate is set free, thereby enriching the sediment in Ca phosphate, and simulating the effect of a high salinity. This method is therefore unusable in sediments, such as the Late Carboniferous of Europe, in which reducing conditions were at one time prevalent. It is additionally inapplicable when the possibility exists that marine strata have been eroded and brought into a milieu of different salinity, as Müller was able to demonstrate. This is, of course, a general prerequisite for almost all geochemical analyses and has already been referred to in connection with boron (chapter 3, "Provenance of the sediment").

Organic substances

In the marine environment the supply of organic substances is considerably richer than in the lakes and rivers of the continents, although since the Devonian organic life in the form of plants has become plentiful on the continents. Correspondingly there should be a clear differentiation in both type and quantity of organic material between sediments of the oceans and the sediments of inland waters. The contrast in type reflects the difference between animal and plant life: the former contain more fats, proteins, and hemin than plants, in which carbohydrate, lignin and chlorophyll are dominant. Both can act as the source of hydrocarbon compounds, nitrogen, and phosphorus.

TREIBS (1934, 1935) was the first to attempt to use organic substances (derivatives of chlorophyll and hemin) to distinguish salinity facies. HUNT et al. (1954) found that an increase in the aromatic carbohydrates with increased proportions of S and N accompanied rising salinity. WELTE (1959) reported limnic organic substances to have a richer variety of molecule spectra than in comparable substances from marine environments. Similarly differences in the C/N ratio may occur, that in marine being less than in limnic sediments. However, this relationship is too easily affected by diagenesis to be of practical worth (ARRHENIUS, 1950; BADER, 1955).

KREJCI-GRAF (1962, 1963) gave the following general description of the behaviour of individual organic substances in sediments. Each oxygen facies is believed to be enriched in different organic substances from those of its neighbours. These substances suffer a different fate during diagenesis. Thus labile protein disappears almost entirely from deposits with easy access for oxygen, whereas in sapropels they remain preserved, although the superficial structure is bacterially altered. Fats are capable of being preserved for a long time in peat-bogs and sapropels, but in better aerated sediments they are quickly lost. Of the carbohydrates cellulose may be enriched in dry conditions but is rapidly decomposed in the presence of water. Bacterial protein and fat may be produced from pectose and nitrogen-free sugars (pentosane). Lignin appears to be absent from plankton and may therefore be characteristic of higher land plants.

"The richness of a rock in organic substances is decided not by their original abundance but by the relation between the organic substances capable of being preserved under the prevalent conditions of deposition and the simultaneously sedimented inorganic substances" (KREJCI-GRAF, 1962, p.7).

The most important organic substances for salinity analysis are apparently the amino acids. They have been studied by many authors the chief of which are ERDMANN et al. (1956), BARGHOORN (1957), SWAIN et al. (1958), ABELSON (1959), BAJOR (1960), DEGENS and BAJOR (1960, 1962), SWAIN (1961), and HARINGTON (1962). Marine sediments appear fundamentally to yield more amino acids than deposits accumulated in fresh water. BAJOR (1960) as well as DEGENS and BAJOR (1962) found not only a greater number of

TABLE X

DISTRIBUTION OF SOME AMINO ACIDS* IN MARINE AND LIMNIC SHALES OF CARBONIFEROUS AGE (After DEGENS and BAJOR, 1962, p.434, 435)

Sample	1	2	3	4	5	6	7	8	9	10	11	12
Marine shale												
Namurian C	3.84	3.33	3.00	5.08	—	2.08	Tr	0.33	4.58	27.80	2.75	2.83
Westphalian A	1.67	2.00	1.67	4.50	—	Tr	Tr	0.50	1.25	9.92	Tr	0.92
Westphalian B	2.50	1.25	3.58	2.00	1.08	2.00	2.83	0.83	3.50	7.32	1.33	5.75
Westphalian C	4.00	4.16	2.25	5.34	Tr	1.42	1.17	0.83	2.08	3.58	19.52	—
									8.92	10.60	2.08	3.58
Limnic shale												
Namurian C	3.25	Tr	2.00	4.34	—	—	3.25	Tr	1.83	—	1.83	2.75
Westphalian A	2.83	—	1.42	6.34	1.17	—	3.96	Tr	1.59	—	0.58	4.00
Westphalian A	5.89	3.90	1.67	1.17	2.22	8.44	8.56	4.10	2.55	Tr	3.25	14.20
Westphalian A	3.00	Tr	2.25	7.42	1.17	Tr	Tr	—	4.08	Tr	1.50	3.08
Westphalian A	2.50	1.87	3.83	3.83	1.75	1.58	5.67	0.50	2.42	1.08	1.58	2.08
Bituminous coal												
Ruhr district	12.05	3.25	8.76	6.25	8.75	12.50	16.02	5.00	10.50	4.25	7.26	33.70
Saar district	15.26	3.25	12.25	5.76	3.25	10.75	15.03	6.25	6.76	7.76	12.77	21.50
Carbonaceous claystone												
Westphalian C	3.78	1.34	2.55	2.00	1.45	2.89	4.89	1.78	5.12	32.50	—	8.88

* 1 = alanine; 2 = tyrosine; 3 = valine; 4 = cystine; 5 = aspartic acid; 6 = glutamic acid; 7 = serine; 8 = threonine; 9 = lysine; 10 = arginine; 11 = proline; 12 = glycine. Contents of amino acids × 10^{-5} %.

amino acids in marine sediments of the Tertiary and Late Carboniferous of western Germany in comparison with limnic horizons, but were able to recognize enrichment in certain particular acids (Table X). Aspartic acid, for instance, is more strongly represented in limnic sediments than in marine, whilst in marine deposits arginine together with lysine are dominant. Arginine has, however, been found in considerable quantity in a carbonaceous clay stone associated with the Erda seam (Ruhr; Westphalian C), which is of nonmarine origin.

A further series of experiments were carried out by the Würzburg Geological Institute on the same Late Carboniferous horizons in the Ruhr used by RICKE (1963) for his sulphur determinations (see p.85) (part of the programme of the International Symposium on the Distinction of Marine and Non-Marine Sediments; see ERNST, 1963). The results of PRASHNOWSKY's (1963) amino-acid investigations given in Table XI show a similar relationship to salinity as outlined above.

Arginine and lysine (diaminomonocarbon acids) were found to form a higher percentage of the amino acids in marine (67.65%) than in limnic sediments (45.85%). However, the reverse relationship is also known: in post-Glacial beds in the Baltic Sea (see also Table XI) arginine and lysine are doubly as rich in limnic diatomaceous sediment (17.88%) as in the marine part of these beds. This is the reverse of the relation in the Late Carboniferous of the Ruhr. These comparisons show that amino acids are by no means a reliable indicator of salinity, the reasons being that they are:

(*1*) Concentrated almost only in fossiliferous and rarely in fossilfree beds (BRIGGS, 1961).

(*2*) Of variable thermal stability; serine, threonine, tyrosine, and the decisively important salinity indicator arginine are, according to VALLENTYNE (1957), CONWAY and LIBBY (1958), particularly unstable.

(*3*) Found in free and combined form. Those present as inorganic or organic complexes are found unchanged as deep as 4,500 m (DEGENS and BAJOR, 1962). However, free amino acids are rarely encountered in fossil sediments.

The occurrence and behaviour of amino acids must therefore be

TABLE XI

DISTRIBUTION OF GROUPS OF AMINO ACIDS * IN SEDIMENTS OF DIFFER-
ENT AGE (After PRASHNOWSKY, 1963, p.304)

Sample	1	2	3	4	5	6	7
Recent muds of the North Sea near Helgoland (*marine*)	23.39	21.45	15.15	26.34	0.86	6.02	6.76
Post-Glacial sediments of the Baltic Sea, *marine* part	22.72	34.82	16.05	8.64	1.28	10.21	6.24
Post-Glacial sediments of the Baltic Sea, *limnic* part	31.95	32.68	5.16	17.88	0.92	9.98	1.40
Bituminous shale of Westphalian A (Ruhr district) *limnic* facies	32.37	17.50	10.91	7.59	19.14	8.81	3.64
Bituminous shale of Westphalian A (Ruhr district) *brackish* facies	45.56	7.34	22.78	—	14.17	10.12	—
Limnic shale Westphalian A	17.90	10.28	11.88	45.85	10.43	3.63	—
Marine shale Namurian C	15.43	3.42	6.20	67.65	4.50	2.61	—
Lower Permian shale Emsland (Germany) saline facies	32.57	12.63	37.33	5.00	3.83	7.80	0.80

* *1* = monoaminomonocarbon acids (alinine, glycine, leucine, isoleucine, valine; *2* = oxymonoaminomonocarbon acids (serine, threonine); *3–5* = aliphatic amino acids: *3* = monoaminodicarbon acids (aspartic and glutamic acid), *4* = diaminomonocarbon acids (arginine, lysine), *5* = aliphatic amino acids with S-contents (cystine, cysteine, methionine); *6* = aromatic amino acids (thyroxine etc.); *7* = heterocyclic amino acids (histidine, tryptophan, proline etc.). Contents of the amino acid groups in %.

further investigated in long continuous profiles.

Similar difficulties arise in dealing with carbohydrates. Vegetable matter when dry may consist of up to 90% carbohydrate comprising monosaccharides (simple sugars), sugar-like polysaccharides, and polysaccharides themselves (starch, cellulose). Pentoses and hexoses are the commonest in nature, but galaktose, glucose and mannose are dominant in sea water. In both Recent and Late Carboniferous sediments glucose, ribose, and rhamnose dominate (Table XII).

TABLE XII

DISTRIBUTION OF CARBOHYDRATES * IN SEDIMENTS OF DIFFERENT AGE
(After PRASHNOWSKY, 1963, p.300)

Sample	1	2	3	4	5	6	7
Recent muds of the North Sea near Helgoland (*marine*)	15.80	11.81	17.85	18.00	18.97	3.94	13.60
Post-Glacial sediments of the Baltic Sea (*marine* part)	11.49	13.76	12.71	18.35	7.58	4.68	31.40
Post-Glacial sediments of the Baltic Sea (*limnic* part)	19.77	7.58	30.77	19.83	9.70	4.53	7.78
Bituminous shale of Westphalian A (*limnic* facies)	17.13	11.15	25.10	17.52	7.47	21.51	—
Bituminous shale of Westphalian A (*brackish* facies)	16.15	13.34	25.28	18.25	1.68	25.28	—
Limnic shale Westphalian A	45.88	9.39	1.18	40.52	0.82	1.51	0.65
Marine shale Namurian C	17.73	27.37	11.30	15.45	5.06	7.35	15.71
Saline shale Lower Permian	27.37	31.70	12.24	10.80	4.89	12.96	—

* *1* = galactose; *2* = glucose; *3* = mannose; *4* = arabinose; *5* = xylose;
6 = Ribose; *7* = rhamnose. Contents of carbohydrates in %.

The preservation and use of these carbohydrates depend essentially on diagenesis, the fossils present and such factors as the transport of continental vegetable material into the sea. Further investigation is also required here.

Oxygen facies

After palaeosalinity the oxygen facies is one of the most popular areas of study in geochemical facies analysis. Under this heading can be understood the demonstration of the original oxygen distribution in sediments, which in turn decides which organic substances may be preserved and which not. The elements C, H, N, and S are employed as well as the heavy metals Ni, Co, V, Cr, (Cu, U). In clastic deposits (relatively coarse grain size, light colour) as opposed to sapropels the greater part of the organic material is destroyed through oxidation. The only organic substances which remain are resin or wax in terrestrial or land proximate sediments. In deposits of nonterrestrial facies scleroproteins and chitin are dominant in which a C/N ratio of between 10 and 12 in recent and 14 and 15 in fossil sediments has been reported (KREJCI-GRAF, 1962, 1963).

In carbonaceous deposits cellulose may be preserved in dry positions, otherwise wax, resins, spores and fungal sclerotia are left. All these substances are durable under a cover of water, but cellulose and lignin, the latter of which is normally lost during decay whereas hemicellulose is scarcely altered, are only partially preserved. Through the preservation of a large quantity of nitrogen-free substances a C/N ratio of between 50 and 100 results in such carbonaceous deposits.

Gyttja is a sediment formed in oxygen-poor water. The presence of O_2 leads to the destruction of protein. Low oxygen concentration allows the preservation of not only such resistant substances as wax and resins but also carbohydrates (pectoses, pentoses). Cellulose is rapidly decomposed in the presence of O_2. Substances rich in N_2 play only a minor part in this milieu, so that the C/N ratio may range between 70 and 350.

Body wax (in recent deposits a mixture of free fatty acids, for instance palmitic and stearic acids, with the Na, Ca, and NH_4 soaps of these fatty acids) is known from the transition zone between a gyttja and a sapropel. Characteristic organic substances of this milieu are porphyrins (according to Krejci-Graf descendants not only of chlorophylls but also hemins).

Characteristic inorganic constituents of gyttja are Co, Cr, (Br and I). The sulphide maximum lies in the boundary zone between gyttja and sapropel. On the gyttja side P, Br, I and (U) are enriched but their maxima, with the exception of P, persist into the sapropel.

A sapropel, although also an organogenetic sediment, is formed under quite different oxidation conditions from gyttja. In the latter the zero level of the redox potential—i.e., the boundary between oxidation and reduction—lies within the sediment itself, whereas in a sapropel it lies above the sediment–water interface. This H_2S protection zone may lie high in the water, as in the Black Sea, so that many otherwise easily decomposed organisms, such as plankton, may be supplied to the sediment in large quantities practically undecayed. Bacterial activity is very intensive in this O_2-free zone and results in the break-down of the organic structure, even chitine and frame-work proteine are partially attacked, but the chemical substance remains unchanged. Phosphorus is set free and enriched in the water above the sediment (SEIBOLD et al., 1958). Nitrogen occurs as N_2, NH_3, or $(CH_3)_3N$, so that the ratio C/N in the sediment is slightly raised in comparison with protein (recent 3/1, and fossil about 30/1).

Trace elements favoured in this O_2-free environment are Cu, Ni, V, and Mo, as in the gyttja porphyrins are present (Table XIII).

After this division of the sediments of the oxygen facies, based essentially on WASMUND (1930) modified from KREJCI-GRAF (1963), some further details of distribution of trace elements in these sediments are given in Table XIV.

Table XV shows the relations between the heavy metals and the oxygen facies in recent marine to brackish water sediments (see also Fig.25–28). Both examples are of the gyttja type. The index elements agree essentially with those listed in Table XIV for O_2-poor sediments. The V/Cr value lies about or below 1. In the second example (Kyrkfjärden) the highest heavy metal concentrations are found not in the centre of the H_2S yielding basin but in the shallower water on the borders of the basin (Fig.25–28).

SEIBOLD et al. (1958) reported V/Cr values of averaging 3.3 from a recent sapropel in the small Malo Jezero Bay on the Island of Mljet off Dubrovnik (Yugoslavia). Ni and Cr were enriched in the approximately 0.12 mm thick dark layers, which alternate with light coloured lime-rich sediments. The water depth was 29 m of which the lower 9 m contain H_2S for the whole of the year. The author's own investigations have shown that the V/Cr ratio in fossil gyttja is higher than the value for the recent sediment in

TABLE XIII

RARE ELEMENTS IN SEDIMENTS [1] (After KREJCI-GRAF, 1962, p.12, 13)

Sample	Facies	NiO	Co_4O_3	Cr_2O_3	V_2O_3	MoO_3
Sandstones, Upper Tertiary (Romania)	limnic-brackish	1–5	1–5	<1	<10	—
Clays, Upper Tertiary (Romania)	brackish-marine	5–10	1–5	1–5	10–50	—
Limestones, Upper Tertiary (Romania)	limnic-brackish-marine	0.5–1	<1–5	—	<10	—
Brown coal, Pliocene (Romania)	terrestrial	10–50	<5–10	<5–50	<10–50	± 10
Brown coal, Eocene (Germany)	terrestrial	<0.5–5	<5–10	<5–5	± 10	—
Bituminous coal, Upper Carboniferous (Germany)		2–50	40	50	<10–10	30
Recent muds (North Sea)	marine	1–5	1–5	1–10	± 10	—
Flysch clay, Upper Eocene (Romania)	marine	10–50	5–10	5–10	10–50	—
Bituminous shale, Lower Cretacous (Germany)	brackish	5–10	1–10	5–10	10–50	—
Oil shales, Eocene (Germany)	limnic	10–80	5–10	10–100	10–50	—
Shales, Lower Devonian (Germany)	marine	5–50	5–10	10–50	10–50	—
Bituminous shale, Lias ε (Germany)	marine	1–50	5–5	5–10	50–100	10
Cupric shale, Upper Permian (Germany)	marine	50	10	5	100–500	10–50
Kukkersit, Ordovician (Estonia)	marine	5	—	—	<10	<10
Asphalt oil, Sarmatian (Austria)		1,000–10,000	50	5–10	50–100	—
Asphalt oil, Oligocene-Miocene (Venezuela)		200–400	5	<5	<500	—

[1] Contents of rare elements × 10^{-5} %.

TABLE XIV

YIELD OF TRACE ELEMENTS IN O_2-POOR AND O_2-FREE SEDIMENTS

Sediment	O_2	Leading elements	Shale
Gyttja	poor	Cr, Co, J (marine), Mn-oxide, V/Cr = 1	oil shale
Transition zone	—	P, Th, U, Th/U < 2, S, Se, Br, J	
Sapropel	free	Cu, Ni, V, Mo, (Pb, Zn), V/Cr = 2–10	oil and cupric shale

TABLE XV

RELATIONS BETWEEN THE HEAVY METALS AND THE OXYGEN FACIES IN
RECENT MARINE TO BRACKISH WATER SEDIMENTS

	Saahich Inlet, Brit. Columbia	Kyrkfjärden, Baltic, near Stockholm (Sweden)
	Author: GROSS (1964)	Authors: LANDERGREN and MANHEIM (1963)
	Varved, diatomaceous clayed silts	Floor muds of variable composition
elements	contents (p.p.m.)	contents (p.p.m.)
Cu	43	54
Zn		130
Ga		24
Ti	2,700	4,300
Pb		31
Th		11
V	66	60
Cr	59	80
Mo		7
U		12
Mn	320	570
Fe	22,000	42,000
Co	7	12
Ni	27	36
V/Cr	1.1	0.75

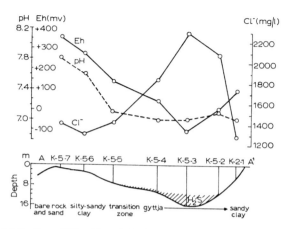

Fig.25. Distribution of Eh, pH and Cl⁻ in the recent sediments of the Kyrkfjord (Sweden). (From LANDERGREN and MANHEIM, 1963, p.177, fig.4; with kind permission of the authors and editor.)

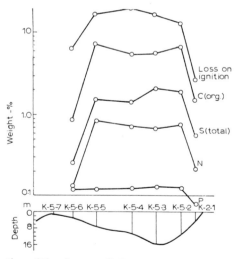

Fig.26. Distribution of the elements C, S, N and P in the Kyrkfjord (Sweden). (From LANDERGREN and MANHEIM, 1963, p.178, fig.6; with kind permission of the authors and editor.)

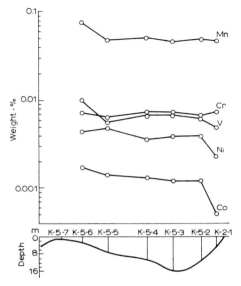

Fig.27. Distribution of Mn, Cr, V, Ni, and Co in the Kyrkfjord (Sweden). (From LANDERGREN and MANHEIM, 1963, p.179, fig.7; with kind permission of the authors and editor.)

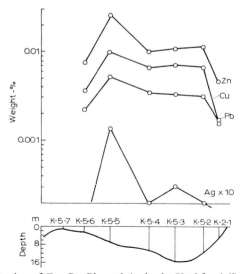

Fig.28. Distribution of Zn, Cu, Pb, and Ag in the Kyrkfjord (Sweden). (From LANDERGREN and MANHEIM, 1963, p.179, fig.8; with kind permission of the authors and editor.)

Table XIV: various systems and localities in Europe, 1.09 (average); Glarner Schiefer (Oligocene), 0.84–2.10; Hunsrück Schiefer (Devonian), 1.33; Lias ε, different locations in Swabia (Germany), 2.30–5.45. The relatively high figures and range of variation of V/Cr in Lias deviate from Table XIV. The grounds may be than in a basin of the size of that in the Early Jurassic of Swabia, O_2-poor and O_2-rich sites may lie beside one another, or, through time, above one another. A further possibility may be found in the grain size dependency of V, which is often richer in the coarser fraction; or in the dependency of V and Cr on the carbonate content of the sediment (BAUMANN, 1968; GITTINGER, 1968). Thus samples to be compared should be of the same order of grain size, and where this is not possible an average value must be obtained by investigating long profiles at several localities, as in the diagnosis of salinity.

The investigation of continuous sequences has the advantage of allowing anamolous analyses to be traced to geological or petrographical peculiarities. In this manner in the exploratory oil boring "Lauben 2" by Memmingen, Allgau (southern Germany), the V/Cr maxima were related to an oil-impregnated sand and silt stone (Bausteinzone, Middle Oligocene). This is the sediment of a well aerated basin, and the high V/Cr ratio cannot be explained by the depositional milieu. However, since the heavy metal content of oil is similar to that of a sapropel (KREJCI-GRAF, 1962) one can assume that the high amount of V in the Baustein beds is traceable to migrating oil. This is an important source of error to be borne in mind in using the V/Cr ratio to diagnose oxygen facies.

Apart from the heavy metal associations already mentioned, the ratios Fe^{2+}/Fe^{3+} and Fe/Mn as well as sulphur may be used in diagnosing oxygen facies. As mentioned in the discussion of palaeosalinity, the sulphur content and the Fe/Mn ratio are dependent on the oxygenation of the depositional environment. The sympathy of the trends of V/Cr and the S content has been shown in Fig.29 (boring "Tettnang 1"). GITTINGER (1968) was able to confirm this association in the Triassic beds of Luxembourg.

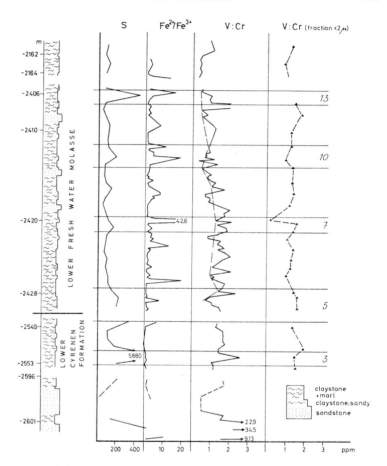

Fig.29. Relationship between the S content and the Fe^{2+}/Fe^{3+} and V/Cr ratios in the lower fresh water molasse of the north Alpine foreland (boring Tettnang 1). Sections *3*, *5*, *7*, *10*, and *13* represent beds with low O_2-potential during deposition. (After BAUMANN, 1968.)

The so-called oil shales are coupled with gyttja and sapropels. BITTERLI (1962, 1963) listed four groups of criteria for differentiating oil shales:

(*1*) Petrographic criteria—mineral composition, texture, colour.

(2) Chemical criteria—content of organic C, N_2, O_2; solubility of organic components and composition of the extracts, trace elements and porphyrins.

(3) Environmental criteria—palaeogeographic position, salinity.

(4) Technical aspect—exploitation and rentability of oil shale.

According to BITTERLI (1963) bituminous substances, in the same form as they appear in oil shales, are found in calcareous, dolomitic, and siliceous clay stones. Bituminous formations appear to be tied to striking geological occurrences (mountain building, transgressions, regressions) which lead indirectly to the creation of closed basins with stagnating conditions, and prolific plankton production.

Among oil shales brackish and chiefly limnic types form the majority, and as a rule have a higher concentration of organic material. Marine oil shales, when they occur, have a much greater extent than those of non-marine origin. Well-known beds of oil shale type in Europe are Lias ε (Posidonien Schiefer: BROCKAMP, 1944; EINSELE and MOSEBACH, 1955; VON GAERTNER and KROEPELIN, 1956), and in the U.S.A. the Green River Formation (Colorado and Utah).

Further important deposits important in the study of oxygen facies are black shales, which in England include marine bituminous shales (MAUCHER, 1962). To black shales are at present generally assigned those beds with more than 2% of organic carbon, which may be present as bitumen or as humic compounds. As we have seen U occurs in the boundary area between gyttja and sapropel. It is enriched in sediments with a high organic content according to KOCZY et al. (1957, 1963) dominantly in the marine domain. STRØM (1948) as well as LANDERGREN and MANHEIM (1963) showed that U achieves high concentrations in fjords and stagnating basins.

For the geochemical behaviour of uranium the redox potential is vital. Salts of U^{6+} are easily soluble in water and can be transported in this form, whilst the quadrivalent form combines under reducing conditions in stagnating water to form complexes with organic substances.

Neither plants nor animals are capable during their lives of concentrating U to the extent in which it can be found in sediments (BREGER and DEUL, 1955). ADAMS and WEAVER (1958) established that in sediments about 20% of the U accompanied by Th is bound to the detritic, acid-insoluble fraction of limestones, dolomites, phosphate rocks, and siliceous shales. The Th content of carbonate rocks is, in addition, dependent on Th-rich resistates (zircon, monazite). In phosphatic rocks in Florida (Bone Valley Formation), the U content is about 250 p.p.m.; in halite, anhydrite, and anhydrite-bearing carbonates U and Th are especially low, whilst in black shales they may each rise to 22–80 p.p.m. In gray and green shales the U content fluctuates between 1 and 12 p.p.m., in sandstones between 0.3 and 1.2 p.p.m.

A Th/U ratio less than 2 characterizes, according to ADAMS and WEAVER (1958), lake and river water, black shales, and phosphate as well as carbonate rocks with detrital silicates. A ratio greater than 7 is indicative of beach sands (monazite) and sediments in which U has been dissolved out through weathering. According to these fairly reliably confirmed values for U and Th and their relationship, U is not exclusively associated with black shales, and is by no means specially enriched in marine sediments. In the enrichment process of U other factors play an important role, for instance the solution of U in ground water and its precipitation in a reducing milieu. This circumstance often arises during diagenesis and is not indicative of the original conditions of deposition. As an index for black shales and oxygen or salinity facies U is therefore to a great extent eliminated.

Sapropels of the cupric shale (Kupferschiefer) type are also of instructive interest. According to RICHTER (1941, p.29) the cupric facies is "clearly restricted to certain zones of deeper water, where the Zechstein limestone also swells to greater thickness". In these zones of deeper water, or in discrete depressions within the Kupferschiefer sea, H_2S is produced and therefore precipitation of heavy metals, dominantly as sulphide occurs, a Co/Ni ratio less or equal to 1; higher V than Cr content; and the antipathetic curves of Zn, Cd and Pb to Cu and Ag indicate the Kupferschiefer as a deposit of an O_2-free facies (KREJCI-GRAZ, 1966). The source of the Cu, the ratio of which to Pb in the Kupferschiefer is much too different from that in sea water (WEDEPOHL, 1964), remains under dispute. To what extent the isotopes of individual elements may be usable indicators of the oxidation facies cannot at the moment be decided. Sediments appear on the whole to have a higher $^{12}C/^{13}C$ ratio than plants (ROSENFELD and SILVERMAN, 1959). In a stagnant milieu as well as in coals, oil shales and oil, ^{12}C is probably concentrated (WICKMANN, 1953), the ^{13}C ratio hardly changing during the conversion to coal or bitumen. Hydrogen isotopes are separated microbiologically (CLOUD, 1965), which may be partially connected with the distribution of O_2 in the sediment. ^{15}N appears to be enriched in anaerobic waters (RICHARDS and BENSON, 1961), and also in amino acids (HOERING and FORD, 1960). ^{32}S is biogenically concentrated but relatively scattered in the sediment (AULT, 1959; RANKAMA, 1963).

Thus for the geochemical analysis of oxygen facies a variety of very different chemical methods is available. A very close relation

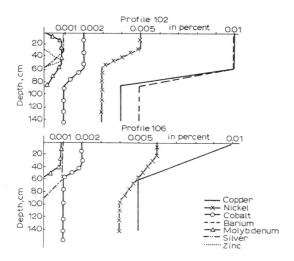

Fig.30. Heavy metals content of loose sediments in the upper part of the Tanalyk River Basin, the southern Urals. (From PAVLENKO and GAVRILOVA, 1964, p.33, fig.13; with kind permission of the editors.)

appears to exist between heavy metals and the oxygen facies, whereby for O_2-poor sediments Cr, Co and Mn oxides, and for O_2-free facies Cu, Ni, V and Mo are diagnostic. However, of these elements Cu and Mo do not have a constant distribution on the Russian

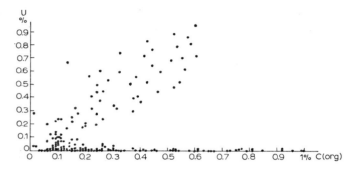

Fig.31. Relation between uranium content (U) and organic-carbon content (C org) in carbonate rocks of the ore-bearing horizon. (From DANCHEV, 1958, p.65, fig.6; with kind permission of the editor.)

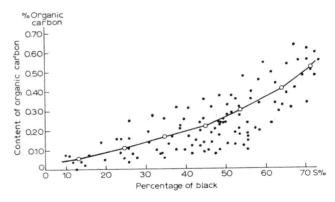

Fig.32. Change in the color of carbonate rocks (limestones and dolomites) in relation to the content of organic carbon. (From DANCHEV, 1958, p.64, fig.5; with kind permission of the editor.)

Platform. In addition Cu, Ni, Co, Mo, as well as Zn may have a quite abnormal distribution in strata laying close to the surface (Fig.30), so that caution is necessary in employing these components in weathering zones. It is to be recommended that as well as these heavy metals the organic components and the colour of the carbonate rocks in relation to the content of organic substances and U be investigated (Fig.31, 32).

For all investigations of oxygen facies the common rule is also valid that statistical investigation of many profiles is necessary before definite conclusions can be drawn.

Temperature facies

The temperature facies is a measure of the original temperature of the water in which sediment was deposited and organisms lived. Of all the types of hydrofacies this is the most difficult to recognize, since the temperature may be quite different at varying levels in the water body, and currents may cause further irregularity. Finally, the original effects of temperature may be overprinted by diagenesis.

Of the elements discussed so far two, B and Br, have been mentioned as dependent on temperature, they follow, however, different behaviour patterns. With regard to boron HARDER (1961) showed that its incorporation in muscovite/illite is dependent on temperature. His experiments find some confirmation in the apparently climatically controlled distribution of the element in the Late Carboniferous, the Early Permian, the Early Cretaceous, and in the Tertiary of Europe (Fig.18). However, a difficulty arises in interpreting the B values as temperature indices since they are also markedly dependent on salinity and the source of the sediment. These factors can only be eliminated in geologically well known areas. In sediments of doubtful source and mode of origin a high boron content may have been derived in a number of ways: (*1*) from boron rich weathering solutions; (*2*) from the immediate neighbourhood of volcanic exhalations; (*3*) evaporation in isolated remnant seas under hot, arid conditions.

In these circumstances only clear, broad-scale tendencies in the distribution of the boron over whole systems can be interpreted as temperature dependent. Absolute temperature can not be read from boron distribution, only relative changes.

Unequivocal temperature data can be obtained from the proportion of Br incorporated in NaCl or KCl minerals. This quantity is strongly dependent on the degree of evaporation of the salt solution and therefore provides a basis for the exact reckoning of temperature (BRAITSCH, 1962), which is unfortunately, however, restricted to saliferous deposits.

As well as the above elements the ratios between Ca/Mg, and Ca/Sr appear to be significant indicators of temperature (CHAVE, 1954; HOLLAND et al., 1963; KINSMAN, 1965). They are, moreover, indicative not only in the mineralogical constituents of the sediment itself but also in fossil shells. Thus the aragonite shell of *Crepidula plana* and those of some other organisms show a positive relationship between the Sr content and the average annual temperature (KREJCI-GRAF, 1966). On the other hand the calcite shell of *Anomia simplex* shows a negative correlation between the two factors (PILKEY and GOODELL, 1964). *Mytilus* shows contrary correlations since in the calcitic prismatic layer there is a positive relationship between Sr content and temperature, whilst in the aragonitic nacreous layer it is negative (DODD, 1964). Connections between the $MgCO_3$ content from the shells of Foraminifera and calcareous algae and the temperature of water are given by Fig.33, 34.

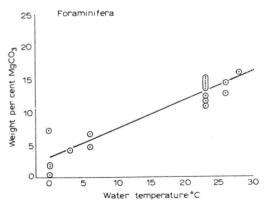

Fig.33. Relation between water temperature and magnesium in the Foraminifera. (From CHAVE, 1954, p. 274, fig.2; with kind permission of the publisher.)

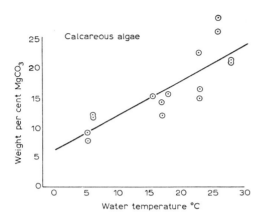

Fig.34. Relation between water temperature and magnesium in the calcareous algae. (From CHAVE, 1954, p.281, fig.14; with kind permission of the publisher.)

Broad-scale differences in temperature between various depositional areas can be concluded from the occurrence of calcium carbonate, which can only be deposited in a warm climate. Finer temperature grading is possible using reefs. Evaporites are also a good indication of a warm climate. The paragenesis of the individual minerals reveals the order of precipitation and depending on water supply and concentration may indicate relative temperatures of deposition. No geochemical investigation is necessary, however, in order to climatically interpret red-earths, red-clay, laterite and bauxite. It is also well known that gypsum, salt and nitrate crusts are produced in a desert climate.

Isotopes have the greatest importance in this field. The most well-known method involves the ratio of the oxygen isotopes $^{16}O/^{18}O$ (VINOGRADOV and TEIS, 1941; UREY, 1947, 1951b; UREY et al., 1948, 1951; EPSTEIN et al., 1951; SILVERMAN, 1951; EMILIANI et al., 1953; TEIS et al., 1957; ENGST, 1961; DEGENS and EPSTEIN, 1964; WEBER, 1964). This method is based on the fact that on the precipitation of calcium carbonate from solution an isotopic differentiation occurs, whereby the ratio of the oxygen isotopes is other than in the water of the solution. This isotopic differentiation is dependent on: (1) temperature, and (2) ratio between the ions in solution.

TABLE XVI

PRECIPITATION TEMPERATURES OF LIMESTONES ACCORDING TO THE ISOTOPE CONCENTRATIONS OF THE OXYGEN (After ENGST, 1961, pp. 149, 162, and the authors cited there)

Sample	$\delta(^{18}O)$		$\delta(^{13}C)$		Temperature (°C)	
	min.	max.	min.	max.	min.	max.
Belemnitella americana, Upper Cretaceous (England)	+0.37	+1.20	−2.20	+0.50	18.4	22.2
Oysters, Upper Cretaceous (England)	−2.05	−0.22	−1.90	−0.20	25.0	34.3
Limestone, Upper Cretaceous (England)	−2.20	0.20	−2.00	+0.10	24.8	35.1
Limestone, Middle Jurassic (Switzerland)	−3.60		−1.30		43.0	
Chalk, Upper Cretaceous (England)	−2.10		−1.20		34.6	
Belemnites mucronata, Cretaceous (N Germany)	+1.79		—		16.8	
Limestone, Cretaceous Maastrichtian (N Germany)	+0.83		—		20.0	
Megateuthis gigantea, Middle Jurassic (N Germany)	+0.16		—		23.1	
Belemnopsis semisulcata, Upper Jurassic (Eichstätt, S Germany)	−0.54		—		26.5	

With the aid of this method the temperature during growth of above all belemnites and oysters has been estimated. Corresponding to the main range of these two groups, palaeotemperature determinations are chiefly available for beds in the Jurassic and Cretaceous. A summary of the most important temperature values is given in Table XVI.

For the application of this method pure uncrystallized calcite is necessary, and therefore one is restricted to using well preserved calcareous fossils. The method is, however, practicable in calcareous shales. The temperature values obtained are, of course, those of the immediate environment of the fossil in question: benthonic organisms can give no information about the temperature other than at the sea floor, and care must be taken to ensure that pelagic fossils bearing information of the temperature at a different level are recognized as such, and also that derived material is not used.

Further isotope measurements in connection with temperature facies have been made using the "sulphate and phosphate thermometer". In view of the high volatility and the possibility of contamination by foreign isotopes, this method is fraught with difficulties (ENGST, 1961).

LITHO- AND BIOFACIES

The litho- and biofacies (see chapter 2 for definitions), which if not independent of the hydrochemical environment are often a relatively insensitive reflection of it, have been more studied geochemically than those aspects of the sediment which are a more direct consequence of the deposition milieu. Such studies have, for instance, dealt with the problem of the early or late diagenetic formation of dolomite, the distribution of elements in reefs, and the dependence of element distribution on palaeogeographic factors (coastal proximity, water depth etc.) from several points of view.

For the theme of this book only those aspects of litho and biofacies which are connected with the hydrofacies are relevant. The reconstruction of palaeogeography stands in first place and is vital in standardizing geochemical results. TOURTELOT (1964) gave a typical example of the extrapolation from lithofacies to the hydrochemical environment. Working along the Late Cretaceous coastline in the western interior of the U.S.A. he was able to recognize several palaeogeographical hydrofacies through the interpretation of sedimentary structures, fossils, and the relationship of beds to one another: non-marine shales, non-marine carbonaceous shales, near-shore marine shales, and off-shore marine shales. Geochemical investigation confirmed the internal consistency of these environments and yielded information valuable in standardizing the geochemical results themselves.

The elements Cr, Mo, Pt, and Zr show no difference between marine and non-marine sediment. Ga is lower in the marine facies than in non-marine samples (see also DEGENS et al., 1957). The further distinction of the divisions of the coastal strip with respect to the trace elements (in p.p.m.) can be seen in Table XVII. In the

TABLE XVII

FURTHER DISTINCTION OF THE ZONES OF THE COASTAL STRIP WITH RESPECT TO THE TRACE ELEMENTS

Zone	Organic C (p.p.m.)	B (p.p.m.)
Non-marine shales	<1	—
Non-marine carbonaceous shales	1–17	99
Near-shore marine shales	<1	112
Off-shore marine shales	circa 8	133

In all zones Co, Mo, Rb, and Sr are approximately the same

As, Cr, V, and Zn increase downwards

near-shore and off-shore zones, B shows no systematic relation to organic C; at best a slight tendency for low B content in samples which are especially rich in organic material is recognizable (c.f., EAGAR, 1962).

V is also independent of organic carbon. The average amount of

V in the near-shore zone is double that in the non-marine carbonaceous shale.

According to VINOGRADOV and RONOV (1956) the concentration of both Al and Ti rises toward the coast. Similarly the Al_2O_3/SiO_2 and the Zr/Hf ratios were stated by RONOV and MIGDISOV (1960) to increase constantly toward the shore in a humid-warm climate. A different horizontal distribution of trace elements is found in large salt basins. In the German Zechstein basin BRAITSCH (1963) found he could differentiate salt facies in the interior of the basin which contained $MgSO_4$, and an $MgSO_4$-poor facies at the borders of the basin. The $MgSO_4$-poor part of the Stassfurt seam corresponds fairly well with the horizontal distribution of the massive Hauptdolomit.

The secondary dolomitization of this originally calcareous bed by $MgSO_4$-bearing sea water is probably the cause of the impoverishment of the salt in this compound. Boron shows a clear rise in content with increasing evaporation and approach to the centre of the basin (Braitsch reported a boron content in the potassic Stassfurt seam of 20 p.p.m. on the margin of the basin and about 400–600 p.p.m. in the centre). Braitsch also found that on the basin margins CaSr-borate accompanied Mg-borate, whereas in the centre of the basin Mg-borate was almost alone. Sr itself is richest on the basin margins where in Zechstein 2 CaSr-borate documents a peripheral facies quite free of $MgSO_4$. The clay mineral distribution is also typical for coast and basin facies of the German Zechstein Basin. According to Braitsch minerals of the chlorite group (corrensite) are dominant on the edge of the basin, whereas talc or even koenenite is formed in the basin centre.

Another type of geochemical investigation concerns reef complexes. In a Late Triassic reef in the Northern Calcareous Alps of Steiermark (Austria) FLÜGEL and FLÜGEL-KAHLER (1963) showed that the amount of $SrCO_3$ increased from the back-reef, and that the $MgCO_3$ content of the reef limestone depended on the type of organism building the frame-work (CHAVE, 1954).

CHAVE was of the opinion that the $MgCO_3$ concentration in different calcareous skeletons is also a question of the water temperature. In echinoderm skeletons the $MgCO_3$ content rose from 5.5 to 12 % in the interval between 5 and 28°C. Identical $MgCO_3$ content in a reef complex may thus indicate a common faunal make-up or a similar environmental temperature. These three examples show that it is quite practicable to pursue in the horizontal dimension the same investigation as in the vertical. A single sediment type may be devided according to the detailed environment of its deposition and in this respect there are further possibilities for investigating dolomite (e.g., DEFFEYES et al., 1964; GITTINGER, 1968), iron ore, and continental rocks.

At the moment knowledge of the genesis of such rocks is insufficient for them to be used as indicators of the environment and they are not yet of great importance in geochemical facies analysis. Finally the biofacies may in certain circumstances be determined by geochemical means. In this field belongs the investigation of trace element distribution in coal. Amongst others the following papers deal with this study in Europe: OTTE (1953), LEUTWEIN and RÖSSLER (1956), JANDA and SCHROLL (1959), BRANDENSTEIN et al.

TABLE XVIII

RARE ELEMENTS (INP.P.M.) IN SAPROPELS, BOGHEAD AND CANNEL COALS OF MIDDLE GERMANY (After LEUTWEIN and RÖSSLER, 1956, p.133)

Element	Sapropels, Öhrenkammer	Boghead coal, Oelsnitz	Boghead-cannel coal, Zwickau
Ge	tr	7	—
Cu	800	17	20
Pb	< 60	210	≤10
Zn	<120	390	40
Ag	1.5	(0.3)	—
As	250	80	(tr)
Sn	—	5	(tr)
Ga	—	120	—
Be	—	—	7
Co	75	265	20
Ni	45	265	20
Mo	<6	tr	<1
V	50	50	20
Mn	(30)	120	—

(1960), and SCHROLL (1961). According to Leutwein and Rössler (see Table XVIII) boghead and cannel coal, formed under O_2-poor conditions, contain a relatively large quantity of Co, Ni, and V. They show a trace element distribution which in part speaks for sapropelic conditions. They are, however, more coals of the gyttja milieu. Anaerobically formed coals such as durain have a similar Co/Ni ratio to boghead and cannel coals.

The individual varieties of coal found segregated in bands yield the following trace elements (after OTTE, 1953):

Vitrain and clarain: Ge, Ga, Be, V, (Ni), Ti, Cu, Mo, Co, Pb, Zn, Sn, Mn.

Durain and fusain: Ti, Cu, Mo, Pb, Zn, Sn, Mn.

Vitrain, clarain and durain can be regarded as anaerobically formed. Fusain is, in contrast, best seen as formed under aerobic conditions. Thus a certain tendency for differential distribution of trace elements appears to exist in coals also. Apart from environmental factors the distribution is also coupled with the ash content of the coal (foreign and plant ash), the type of the coal forming plants, and especially the source of the trace elements.

CHAPTER 6

Non-Chemical Methods of Facies Analysis

In this chapter those possibilities for facies differentiation are given which are quintessentially based on the transformation or neoformation of clay minerals. Such transformations or neoformations occur chiefly under extreme circumstances, e.g., on the transport of clay minerals from a terrestrial into an extreme saline milieu. Changes of this kind were discussed in the previous chapter in connection with the formation of clay minerals in the Zechstein basin.

The transition from montmorillonite via corrensite to talc with increasing salinity is a typical example of how terrestrial clay detritus may be altered as a consequence of a high salt content in the water.

FÜCHTBAUER and GOLDSCHMIDT (1959) were of the opinion that synsedimentary or diagenetic influences could be manifested through:

(1) Environmentally determined clay minerals, if the introduction of detrital material can be excluded.

(2) A difference in the clay mineral type in rocks of the same facies.

(3) Differences in the clay mineral type in alternating beds of different type, for which a change in the source of detritus can be ruled out.

According to the same authors chlorite as a neo-formation is present in small quantities in salt clays accompanied by muscovite, magnesite, haematite, quartz and feldspar. In carbonate rocks muscovite, dolomite or calcite, and pyrite are the normal minerals. In anhydritic rocks accessory constituents are normally synsedimentary limestone as well as muscovite and chlorite (synsedimentary or early diagenetic), montmorillonite, and also mixed-layer minerals (synsedimentary to early diagenetic), magnesite, dolomite and pyrite.

X-ray studies in conjunction with boron analyses in the Late Carboniferous and Early Permian of northwest Germany have shown that the kaolinite there is constantly associated with terrestrial coal-yielding beds, and that beds with illite are marine (STADLER, 1963). Similar observations were made by KRUMM (1963) in the Rhaeto-Liassic beds of east Franconia (Germany) and by KÜBLER (1963) in the northern Sahara.

MILLOT (1949) and WEAVER (1958) were both aware of these relationships and did not rule out the formation of illite during diagenesis. However, today it is recognized that the distribution of clay minerals is less an indication of palaeosalinity than a hint of the weathering and the rock type outcropping in the hinterland (CORRENS, 1963).

FÜCHTBAUER (1963) working in the Tertiary molasse north of the Bavarian Alps found a relationship between the colour of the biotite and tourmaline and the environment of deposition: olive tourmaline occurring in brackish marine sediments, whilst in the limno-fluviatile sediments of the same age only green and brown tourmaline were formed. In the Middle Buntsandstein of western Germany FÜCHTBAUER (1966) observed a connection between the regional distribution of the cementing mineral and the palaeosalinity: of the secondary feldspars albite or plagioclase with albite margins is concentrated in the brackish to marine beds, whilst in the fluviatile area in southern Germany potassic feldspar is dominant. A similar distribution was believed by FÜCHTBAUER (1966) to be present in the Muschelkalk where in the transgressive sediments of the lower Muschelkalk potassic feldspar is present whereas in the beds marking the acme of marine conditions it is replaced by albite.

Secondary quartz is also apparently milieu-dependent in the Middle Buntsandstein, being richer in the estuarine deposits. Anhydrite is the precipitate in sandstones outside the fluviatile zone.

It remains to determine if the secondary changes in the shallow burial stage occur under the influence of pore water which has still a degree of chemical similarity to the original sea water. SCHUSTER (1968), using electrical logging of boreholes, was able to recognize marine and non-marine beds at depths of 100–3,500 m through the electrolytic properties of the pore water. It seems, therefore, that at least in some circumstances the pore water remains as a potent influence through and even beyond the shallow burial stage.

Another possibility for analysing the original environment of deposition is provided by the occurrence of subaquatic shrinkage cracks in the sediment. These shrinkage cracks are often opened from beneath and filled with the underlying sand. According to BURST (1962) such cracks may be originate through the removal of water from a clay with a small quantity of minerals capable of swelling and contracting, if the salt content in the overlying water increases. A further hint of the original water of the area of deposition can be obtained from the compression of the various clay minerals. Thus WHITEHOUSE et al. (1960) reported that kaolinite, illite, and montmorillonite have different rates of compression dependent on the pH and the chlorinity of the water. The pore volume can also provide information over the original hydrofacies (HELING, 1967b). Above all the grain size, the type of the clay minerals, and the salinity of the sediment liquor are important for the early porosity of a clay grade rock, and the porosity of the freshly deposited sediment also rises with increasing electrolyte content.

The original porosity of the clay sediment can be extrapolated by examining the porosities of contemporaneous deposits from different depths of burial. The extrapolated value for the porosity is then multiplied by the natural logarithm of the grain size producing a measure of the original salinity through its relation with this "modified original porosity".

CHAPTER 7

Practical Use and Future Importance of Geochemical Facies Analysis

The practical use of geochemical facies analysis is to be found above all in the correlation of beds and prospection for mineral resources. The most important field is the division of fossil-free sediments, such beds forming a considerable part of the stratigraphical column and often being of economic importance (e.g., the Early Permian gas reservoir at Slochteren near Groningen in The Netherlands). Examples of the course of boron curves in the Early Permian have already been mentioned in the section on boron. The basin sediments of this age in Emsland (northwestern Germany) are red-brown clay and siltstones, which in the interior of the basin are intercalated with a few relatively thin sandstones and even in some places by spilites. Apart from a few spores, *Estheria* and carbonaceous plant remains these beds contain no further characteristic fossils. The stratigraphic position of considerable thickness of beds can be approximately obtained from the order of magnitude of the boron content. It is much less in the lower Rotliegendes than in the upper part. A finer subdivision is not possible since local unconformities (caused by local erosion), increase of the sand component, and differing rates of evaporation of the remnant seas lacking drainage caused considerable variation of the boron content within narrow intervals.

The division of beds with the aid of boron is more successful in the uniformly deposited, more or less mineralogically homogenous clay stones of the sub-Variscan fore-deep, or in the quite depositional sites in the basin facies of the fore-Alpine molasse of southern Germany. How far the potentiality for division and correlation rests on the relation of boron to variation of a single facies phenomenon and not several is, however, still uncertain.

The use of Br for the division of marine salt deposits is more reliable. This element is incorporated in sylvine (KCl) and halite (NaCl) in place of chlorine as the degree of evaporation of the saline liquor increases, and it therefore documents the temporal succession of the precipitates which corresponding to the same physico-chemical conditions can be correlated within the same basin over considerable distances. The method fails in the case of salt deposits affected by high grade metamorphism, but does then allow the degree of alteration of the salt to be estimated.

Relative dating can also be carried out using the ratios of the heavy metals in O_2-poor sediments, especially Fe^{2+}/Fe^{3+}, Fe/Mn and V/Cr, which can also be correlated between themselves. The division of such sediments is, however, dependent on their constitution (argillaceous or calcareous) and on the grain size. A high V/Cr ratio may follow on the impregnation of beds with oil and this is not to be interpreted as an indication of the facies. In borings it is possible during the boring itself to anticipate the cutting of an oil-impregnated bed by observing a rise in the V/Cr ratio in aerobic beds caused by the high V content of the migrated oil.

For a long time the so-called "oil shales" have been geochemically investigated in an attempt to trace the source rock of migrated oil. The oil shales, deposited under oxygen-poor or free conditions have been investigated from the very different points of view of many authors: POTONIE (1928), TRASK (1932), VON GAERTNER and KROEPELIN (1956), WEBER (1958), HECHT (1959), JACOB (1961), BITTERLI (1962, 1963). These investigations have shown that although the formation of oil can be confidently traced to sapropels it is not usually possible to nominate a particular source rock from knowledge of the chemical composition of the oil.

The exact geochemical facies analysis of coals yields information on the possibility of satisfactorily producing briquets, the combustion properties, and the accessory products which can be expected during coking. Marine influence in the coarse brown coal of the lower Rhine near Cologne has led to difficulty in making briquets and to unsatisfactory burning. In particular their stability when used as fuel is poor (HAASE and PFLUG, 1958) and much slag is produced in the furnace. These disadvantageous features are caused by a low pore volume, low plasticity, as well as a high ash content. Moreover the content of organically bound S and Al_2O_3 are considerably higher than in those coals which were formed under limnic conditions.

The facies of the coal also appears to be connected with the origin or retention of gases of the most varied composition. Coal layers which geochemical investigation reveals as formed under oxygen-free conditions (sapropel coals) contain a high proportion of gaseous hydrocarbons. Similarly preliminary investigations made by REINKENSMEIER (1970) indicate that bituminous clay-stones can also be distinguished through their gas spectra. The gases developed in the course of diagenesis remain in part bound to the organic material or clay minerals with a relatively large surface area and can be released by desorption analysis (grinding of the sample in gas-proof containers followed by heating). Those gases which in contrast are freed during diagenesis wander through succeeding permeable beds (through fine hair-cracks and fractures) to the surface where it can be collected in shallow bore holes. The first measurements of this kind made by ERNST (1968) have shown that the position, depth and dip of such gas releasing bituminous shales of the gyttja or sapropel type can be clearly delimited on the surface, providing that the bed is not deeper than 30 m. This type of subsurface mapping is especially important in regions hidden under a cover of Pleistocene deposits.

The future importance of geochemical facies analysis also lies in the investigation of sediments lacking distinctive fossils or minerals. Their division into facies can, as described above, be made using those elements with a steady concentration in the oceans over a long period, and which are relatively insensitive to weathering and diagenesis as well as being easily chemically determinable. Few elements are in fact suitable, and among them are the rare earths which may be of value for salinity and oxygen facies, but which have not yet been by any means thoroughly investigated in this regard.

Further possibilities lie in determining the water depth of a basin, for which Br may be used in marine salt deposits, and otherwise isotopes of oxygen, phosphor and nitrogen.

Important new results can, above all, be expected from the investigation of organic substances and organic clays. This field includes the investigation of amines, amino acids, carbohydrates, phenols, hydrocarbons, alcohols and enzymes associated with such complexes (DEGENS, 1968; and the authors listed there), as well as the interdigitation or adsorption of metals in organic complexes.

Investigation of this kind require close cooperation between geologists, palaeontologists, mineralogists, petrographers, chemicals and botanists. It is also very closely connected with those studies in the geosciences and molecular chemistry which are concerned with the origin and distribution of organic life on earth. In this field geochemical facies analysis has its own contribution to make.

References

ABELSON, P. H., 1959. Geochemistry of organic substances. In: P. H. ABELSON (Editor), *Researches in Geochemistry*. Wiley, New York, N.Y., 511 pp.

ADAMS, J. and WEAVER, C. E., 1958. Thorium/uranium ratios as indicators of sedimentary processes: example of concept of geochemical facies. *Bull. Am. Assoc. Petrol. Geologists*, 42 (2): 387–430.

ARRHENIUS, G., 1950. Carbon and nitrogen in subaquatic sediments. *Geochim. Cosmochim. Acta*, 1: 15.

AULT, W. U., 1959. Isotopic fractionation of sulfur in geochemical processes. In: P. H. ABELSON (Editor), *Researches in Geochemistry*. Wiley, New York, N.Y., 511 pp.

AULT, W. U. and KULP, J. L., 1959. Isotopic geochemistry of sulphur. *Geochim. Cosmochim. Acta*, 16: 201–235.

BADER, R. G., 1955. Carbon and nitrogen relations in surface and sub-surface marine sediments. *Geochim. Cosmochim. Acta*, 7: 205.

BAJOR, M., 1960. Amine, Aminosäuren und Fette als Faziesindikatoren. *Braunkohle*, 12: 472–478.

BARGHOORN, E. S., 1957. Origin of life—treatise on marine ecology and paleoecology, 2. *Geol. Soc. Am., Mem.*, 67: 75–86.

BARTH, T., CORRENS, C. und ESKOLA, P., 1939. *Die Entstehung der Gesteine*. Springer, Berlin, 422 pp.

BAUMANN, A., 1968. *Chemisch-geologische Faziesuntersuchungen in der Molasse des Nördlichen Bodenseegebietes*. Dissertation, Univ. Tübingen, 121 pp.

BEHNE, W., 1953. Untersuchungen zur Geochemie des Chlor und Brom. *Geochim. Cosmochim. Acta*, 3: 186–215.

BISCHOF, G., 1847–51. *Lehrbuch der Chemischen und Physikalischen Geologie, 1–2*. Marcus, Bonn, 989/844 pp.

BITTERLI, P., 1962. Untersuchung bituminöser Gesteine von West-Europa. *Erdöl Kohle*, 15: 2–6.

BITTERLI, P., 1963. Classification of bituminous rocks of western Europe. *World Petrol. Congr., Proc., 6th, Frankfurt/Main, 1963, Sect. I*, 30: 11 pp.

BLUMER, M., 1950. Porphyrinfarbstoffe und Porphyrin-Metallkomplexe in schweizerischen Bitumina. *Helv. Chim. Acta*, 33: 1627–1637.

BLUNDELL, C. R. K. and MOORE, L. R., 1960. Mid-coal measures "red beds" in

---

[1] All quotations of Russian literature after V. P. MARKEVICH, 1960. *The Concept of "Facies"*, 1–3. *Intern. Geol. Rev., Washington*, 2.

the south Wales coalfield. *Congr. Avan. Études Stratigraph. Géol. Carbonifère, Compt. Rend., 4, Heerlen, 1958*, 1: 41–48.

BOEKE, H. und EITEL, W., 1923. *Die Grundlagen der Physikalisch-chemischen Petrographie*, 11, 2 Aufl. Bornträger, Berlin, 589 pp.

BRADACS, L. K. und ERNST, W., 1956. Geochemische Korrelationen im Steinkohlenbergbau. *Naturwissenschaften*, 43: 33.

BRAITSCH, O., 1962. *Mineralogie und Petrographie in Einzeldarstellungen, 3. Enstehung und Stoffbestand der Salzlagerstätten*. Springer, Berlin–Heidelberg–Göttingen, 232 pp.

BRAITSCH, O., 1963. Evaporite aus normalem und verändertem Meerwasser. *Fortschr. Geol. Rheinland Westfalen*, 10: 151–172.

BRANDENSTEIN, M., JANDA, I. und SCHROLL, E., 1960. Seltene Elemente in österreichischen Kohlen- und Bitumengesteinen. *Mineral. Petrog. Mitt.*, 7 (3): 260–285.

BREGER, J. A. and DEUL, M., 1955. The organic geochemistry of uranium. *U.S., Geol. Surv., Proffess. Papers*, 300: 505–510.

BRIGGS, M. H., 1961. Amino acids and peptides from some New Zealand fossils. *New Zealand J. Geol. Geophys.*, 4: 387–391.

BRINKMANN, R. und DEGENS, E., 1956. Die geochemische Verteilung einiger Elemente im Ruhrkarbon. *Naturwissenschaften*, 43: 56.

BROCKAMP, B., 1944. Zur Paläogeographie und Bitumenführung des Posidonienschiefers im deutschen Lias. *Arch. Lagerstättenforschung*, 77: 59 pp.

BUGRY, R., REYNOLDS JR., R. C. and SHAW, D. M., 1964. Unreliable boron analyses in palaeo–salinity investigations. *Nature*, 201: 1314–1316.

BURST, J. F., 1958a. "Glauconite" pellets: their mineral nature and applications to stratigraphic interpretations. *Bull. Am. Assoc. Petrol. Geologists*, 42: 310–327.

BURST, J. F., 1958b. Mineral heterogeneity in "glauconite" pellets. *Am Mineralogist*, 43: 481–497.

BURST, J. F., 1962. Subaqueously formed shrinkage cracks in clay. *J. Sediment. Petrol.*, 35: 348–353.

CAMERON, E. M., 1966. A geochemical method of correlation for the lower cretaceous strata of Alberta. *Geol. Surv. Can., Bull.*, 137: 29 pp.

CARSTENS, H., 1949. Et nytt prinsipp ved geochemiske beregninger. *Norsk. Geol. Tidsskr.*, 28: 47–50.

CHAVE, K. E., 1954. Aspects of the biochemistry of magnesium, 1. Calcareous marine organisms. *J. Geol.*, 62: 266–283.

CHILINGAR, G. V., 1953. Use of Ca/Mg ratio in limestones as a geological tool. *Compass*, 30: 202–209.

CLARKE, F. W., 1924. *The Data of Geochemistry*, 5th ed. — *U.S., Geol. Surv., Bull.*, 770: 841 pp.

CLAYTON, R. N., 1959. Oxygen isotope fractionation in the system calcium-carbonate–water. *J. Chem. Phys.*, 30: 1246–1250.

CLAYTON, R. N. and DEGENS, E. T., 1959. Use of carbon isotope analyses of carbonates for differentiation of fresh water and marine sediments. *Bull. Am. Assoc. Petrol. Geologists*, 43: 890–897.

CLOUD, P. E., 1965. Significance of the Gunflint (Precambrium) microflora. *Science*, 148: 27–35.

CLOUD, W. H., 1952. The 15-micron Band of CO_2 Broadened by Nitrogen and Helium. O.N.R. Program, Rept., John Hopkins Univ., Baltimore.

CONWAY, E. J., 1942. Mean geological data in relation to oceanic evolution. *Proc. Roy. Irish Acad., Sect. B*, 48: 119–159.

CONWAY, E. J., 1943. The chemical evolution of the ocean. *Proc. Roy. Irish Acad., Sect. B*, 48: 161–212.

CONWAY, D. and LIBBY, W. F., 1958. The measurement of very slow reaction rates, decarboxylation of alanine. *J. Am. Chem. Soc.*, 80: 1077–1084.

CORRENS, C. W., 1948. Die geochemische Bilanz. *Naturwissenschaften*, 35: 7–12,

CORRENS, C. W., 1956. The geochemistry of the halogens. In: L. H. AHRENS. F. PRESS, K. RANKAMA and S. K. RUNCORN (Editors), *Physics and Chemistry of the Earth*, 1. Pergamon, London, 317 pp.

CORRENS, C. W., 1963. Tonminerale. *Fortschr. Geol. Rheinland Westfalen*, 10: 307–318.

CURTIS, C. D., 1964. Studies of the use of boron as a paleo-environmental indicator. *Geochim. Cosmochim. Acta*, 28: 1125–1137.

DALY, R. A., 1907. The limeless ocean of Precambrian time. *Am. J. Sci.*, 23: 93–115.

DANCHEV, V. J., 1958. The importance of quantitative determination of color in the study of sedimentary uranium deposits. *Bull. Acad. Sci. U.S.S.R., Geol. Ser., English Transl.*, 1958: 60–71.

D'ANS, J. und KÜHN, R., 1940. Über den Bromgehalt von Salzgesteinen der Kalisalzlagerstätten, 1–5. *Kali*, 34: 42–46, 59–64, 77–83.

D'ANS, J. und KÜHN, R., 1944. Über den Bromgehalt von Salzgesteinen der Kalisalzlagerstätten, 6. *Kali*, 38: 167–169.

DEFFEYES, K. S., LUCIA, F. J. and WEYL, P. K., 1964. Dolomitization: observations on the island of Bonaire, Netherland Antilles. *Science*, 143: 678–679.

DEGENS, E. T., 1958a. Geochemische Untersuchungen zur Faziesbestimmung im Ruhrkarbon und im Saarkarbon. *Glückauf*, 94: 513–520.

DEGENS, E. T., 1958b. Environmental studies of Carboniferous sediments, 2. Application of geochemical criteria. *Bull. Am. Assoc. Petrol. Geologists*, 42: 981–987.

DEGENS, E. T., 1968. *Geochemie der Sedimente*. Enke, Stuttgart, 282 pp.

DEGENS, E. T. und BAJOR, M., 1960. Die Verteilung von Aminosäuren in bituminösen Sedimenten und ihre Bedeutung für die Kohlen- und Erdölgeologie. *Glückauf*, 96: 1525–1534.

DEGENS, E. T. und BAJOR, M., 1962. Die Verteilung der Aminosäuren in limnischen und marinen Schiefertonen des Ruhrkarbons. *Fortschr. Geol. Rheinland Westfalen*, 3(2): 429–440.

DEGENS, E. T. and CHILINGAR, G. V., 1967. Diagenesis of subsurface waters. In: G. LARSEN and G. V. CHILINGAR (Editors), *Diagenesis in Sediments*. Elsevier, Amsterdam, pp.477–502.

DEGENS, E. T. and EPSTEIN, S., 1964. Oxygen and carbon isotope ratios in coexisting calcites and dolomites from recent and ancient sediments. *Geochim. Cosmochim. Acta*, 28: 23–44.

DEGENS, E. T., WILLIAMS, E. G. and KEITH, M. L., 1957. Environmental studies of Carboniferous sediments, 1. Geochemical criteria for differentiating marine and fresh-water shales. *Bull. Am. Assoc. Petrol. Geologists*, 41: 2427.

DEGENS, E. T., EMERY, O. and REUTER, J. H., 1963. Organic materials in recent and ancient sediments, 3. Biochemical compounds in the San Diego Trough, California. *Neues Jahrb. Paläontol., Monatsh.*, 1963: 231–248.

DODD, J. R., 1964. Environmentally controlled strontium and magnesium variation in *Mytilus*. *Geol. Soc. Am., Abstr.*, 1963: 46.

DUNBAR, C. O. and RODGERS, J., 1957. *Principles of Stratigraphy*. Wiley, New York, N.Y., 356 pp.

DUNNING, H. N., MOORE, J. W. and MYERS, A. T., 1954. Properties of porphyrins in petroleum. *Ind. Eng. Chem.*, 46: 2000–2007.

EAGAR, R. M. C., 1962. Boron content in relation to organic carbon in certain sediments of the British coal measures. *Nature*, 196: 428–431.

ECKHARDT, F. J., 1958. Über Chlorite in Sedimenten. *Geol. Jahrb.*, 75: 437–474.

EINSELE, G. und MOSEBACH, R., 1955. Zur Petrographie, Fossilerhaltung und Entstehung der Gesteine des Posidonienschiefers im Schwäbischen Jura. *Neues Jahrb. Geol. Paläontol., Abhandl.*, 101: 319.

EMERY, K. O. and RITTENBERG, S. C., 1952. Early diagenesis of California Basin sediments in relation to origin of oil. *Bull. Am. Assoc. Petrol. Geologists*, 36: 735–806.

EMILIANI, C., EPSTEIN, S. and UREY, H. C., 1953. Temperature variations in the Lower Pleistocene of southern California. *J. Geol.*, 61: 171–181.

ENGST, H., 1961. *Über die Isotopenhäufigkeiten des Sauerstoffes und die Meerestemperatur im süddeutschen Malm-Delta*. Dissertation, J. W. Goethe Univ., Frankfurt/Main, 184 pp.

EPSTEIN, S., BUCHSBAUM, R., LOWENSTAM, H. and UREY, H. C., 1951. Carbonate–water isotopic temperature scale. *Bull. Geol. Soc. Am.*, 62: 417–425.

ERDMANN, J. G., MARLETT, E. M. and HANSON, W. E., 1956. Survival of amino acids in marine sediments. *Science*, 124: 1026.

ERNST, W., 1962. Die fazielle und stratigraphische Bedeutung der Borgehalte im jüngeren Oberkarbon und Rotliegenden Nordwestdeutschlands. *Fortschr. Geol. Rheinland Westfalen*, 3 (2): 423–428.

ERNST, W., 1963. Beschreibung und Vergleich von Analysendaten der Gemeinschaftsuntersuchung unter geologischen Aspekten. *Fortschr. Geol. Rheinland Westfalen*, 10: 347–362.

ERNST, W., 1964. *Beitrag der Bor-Methode zur Paläogeographie des Nordwestdeutschen Oberkarbons und Unterperms*. Habilitationsschrift, Tübingen, 60 pp.

ERNST, W., 1966. Stratigraphisch-fazielle Identifizierung von Sedimenten auf chemisch-geologischem Wege. *Geol. Rundschau*, 55: 21–29.

ERNST, W., 1968. Verteilung und Herkunft von Bodengasen in einigen süddeutschen Störungszonen. *Erdöl Kohle*, 21: 605–610, 692–697.

ERNST, W., 1969. Nachweis der Erdgezeiten mit Bodengasen. *Meteorol. Rundschau*, 22(5): 140–142.

ERNST, W. und LODEMANN, W., 1965. Die Verteilung des Bors im Kristallin der SE-Saualpe/Ostkärnten. *Neues Jahrb. Geol. Paläontol., Monatsh.*, 11: 641–647.

ERNST, W. und WERNER, H., 1960. Die Bestimmung der Salinitätsfazies mit Hilfe der Bor-Methode. *Glückauf*, 96: 1064–1070.

ERNST, W., KREJCI-GRAF, K. und WERNER, H., 1958. Parallelisierung von Leithorizonten im Ruhrkarbon mit Hilfe des Bor-Gehaltes. *Geochim. Cosmochim. Acta*, 14: 211–222.

FAIRBRIDGE, R. W., 1967. Phases of diagenesis and authogenesis. In: G. LARSEN and G. V. CHILINGAR (Editors), *Diagenesis in Sediments*. Elsevier, Amsterdam, pp.19–89.

FLÜGEL, E. und FLÜGEL-KAHLER, E., 1963. Mikrofazielle und geochemische Gliederung eines obertriadischen Riffes der nördlichen Kalkalpen (Sauwand bei Gusswerk, Steiermark, Österreich). *Mitt. Mus. Bergbau, Geol. Technol.*, 24: 1–129.

FREDERICKSON, A. F. and REYNOLDS, R. C., 1960. Geochemical method for determining palaeosalinity. *Clays Clay Minerals, Proc. Natl. Conf. Clays Clay Minerals*, 8 (1959): 201–213.

FÜCHTBAUER, H., 1956. Zur Entstehung und Optik authigener Feldspäte. *Neues Jahrb. Mineral., Monatsh.*, 1956: 9–23.

FÜCHTBAUER, H., 1963. Zum Einfluss des Ablagerungsmilieus auf die Farbe von Biotiten und Turmalinen. *Fortschr. Geol. Rheinland Westfalen*, 10: 331–336.

FÜCHTBAUER, H., 1966. Der Einfluss des Ablagerungsmilieus auf die Sandsteindiagenese im mittleren Buntsandstein. *Sediment. Geol.*, 1: 159–179.

FÜCHTBAUER, H., 1967. Die Sandsteine in der Molasse nördlich der Alpen. *Geol. Rundschau*, 56: 266–300.

FÜCHTBAUER, H. und GOLDSCHMIDT, H., 1959. Die Tonminerale der Zechsteinformation. *Beitr. Mineral. Petrogr.*, 6: 320–345.

FÜCHTBAUER, H. und GOLDSCHMIDT, H., 1963. Beobachtungen zur Tonmineral-Diagenese. *Proc. Intern. Conf. Clay, 1st, Stockholm, 1963*, pp.99–111.

GITTINGER, K., 1968. *Geochemische Faziesuntersuchungen im Oberen Hauptmuschelkalk und Unteren Keuper Luxemburgs*. Dissertation. Univ. Tübingen, 123 pp.

GOLDBERG, E. D. and ARRHENIUS, G. O. S., 1958. Chemistry of Pacific pelagic sediments. *Geochim. Cosmochim. Acta*, 13: 153–212.

GOLDSCHMIDT, V. M., 1933. Grundlagen der quantitativen Geochemie. *Fortschr. Mineral.*, 17: 112–156.

GOLDSCHMIDT, V. M., 1954. *Geochemistry*. Oxford Univ. Press, London, 730 pp.

GOLDSCHMIDT, V. M. und PETERS, C. L., 1932. Zur Geologie des Bors. *Nachr. Akad. Wiss. Göttingen, Math. Physik. Kl. IIa*, 3: 402.

GOLOVKINSKII, N. A., 1869. *O Permskoi Formatsii v Tsentralnoi Chasti Kamsko-Volzhskogo Basseina, Materialy dlia Geologii Rossii, 1*. (About Permian formation in the central part of Kama-Volga Pasin, 1).

GREGOR, C. B., 1967. The geochemical behaviour of sodium with special reference to post-Algonkian sedimentation. *Verhandel. Koninkl. Ned. Akad. Wetenschap., Afdel. Natuurk., Sect. 1*, 24 (2): 66 pp.

GRIM, R. E. and JOHNS, W. D., 1954. Clay-mineral investigation of sediments in the northern Gulf of Mexico. *Clays Clay Minerals, Proc. Natl. Conf. Clays Clay Minerals*, 2 (1954): 81–103.

GRIM, R. E. and LOUGHNAN, F. C., 1962. Clay minerals in sediments from Sydney Harbour, Australia. *J. Sediment. Petrol.*, 32: 240–248.

GRIM, R. E., DIETZ, R. S. and BRADLEY, W. F., 1949. Clay-mineral composition of some sediments from the Pacific Ocean of the California Coast and the Gulf of California. *Bull. Geol. Soc. Am.*, 60: 1785–1808.

GROENNINGS, S., 1953. Quantitative determination of the porphyrin aggregate in petroleum. *Anal. Chem.*, 25: 938–941.

GROSS, M. G., 1964. Heavy-metal concentrations of diatomaceous sediments in a stagnant fjord. *Geol. Soc. Am.*, *Spec. Papers*, 76: 69.

HAASE, F. und PFLUG, H. D., 1958. Fazies und Brikettierbarkeit der niederrheinischen Braunkohle. *Fortschr. Geol. Rheinland Westfalen*, 2: 613–632.

HARDER, H., 1959. Beitrag zur Geochemie des Bors. *Fortschr. Mineral.*, 37 (1): 82–87.

HARDER, H., 1961. Einbau von Bor in detritische Tonminerale. *Fortschr. Mineral.*, 39 (1): 148–149.

HARDER, H., 1963. Inwieweit ist das Bor ein marines Leitelement? *Fortschr. Geol. Rheinland Westfalen*, 10: 239–252.

HARE, P. E., 1962. *The amino acid composition of the organic matrix of some recent and fossil shells of some west coast species of Mytilus*. Thesis, Calif. Inst. Technol., Div. Geol. Sci.

HARINGTON, J. S., 1962. Natural occurrence of amino acids in Virgin crocidolite asbestos and banded ironstone. *Science*, 138: 521–522.

HARRISON, A. G. and THODE, H. G., 1958a. Sulphur isotope abundances in hydrocarbons and source rocks of Uinta Basin, Utah. *Bull. Am. Assoc. Petrol. Geologists*, 42: 2642–2649.

HARRISON, A. G. and THODE, H. G., 1958b. Mechanism of the bacterial reduction of sulphate from isotope fractionation studies. *Trans. Faraday Soc.*, 54:84–92.

HECHT, F., 1959. Migration, Tektonik und Erdöllagerstätten im Gifhorner Trog. *Erdöl Kohle*, 12: 303.

HECHT, F., 1963. Uran- und Thoriumbestimmungen in österreichischen Wässern und Gesteinen. *Fortschr. Geol. Rheinland Westfalen*, 10: 193–200.

HEDBERG, H. D., 1926. The effect of gravitational compaction on the structure of sedimentary rocks. *Bull. Am. Assoc. Petrol. Geologists*, 10: 1035.

HELING, D., 1967a. Die Salinitätsfazies von Keupersedimenten aufgrund von Borgehaltsbestimmungen. *Sedimentology*, 8: 63–72.

HELING, D., 1967b. Die Porositäten toniger Keuper- und Jura-Sedimente Südwestdeutschlands. *Contr. Mineral. Petrol.*, 15: 224–232.

HELLER, W., 1965. Organisch-chemische Untersuchungen im Posidonienschiefer Schwabens. *Neues Jahrb. Geol. Paläontol.*, *Monatsh.*, 1965: 65–68.

HIRST, D. M., 1962a. The geochemistry of modern sediments from the Gulf of Paria, 1. The relationship between the mineralogy and the distribution of major elements. *Geochim. Cosmochim. Acta*, 26: 309–334.

HIRST, D. M., 1962b. The geochemistry of modern sediments from the Gulf of Paria, 2. The location and distribution of trace elements. *Geochim. Cosmochim. Acta*, 26: 1147–1187.

HOERING, TH. C. and FORD, H. T., 1960. The isotope effect in the fixation of nitrogen by azotobacter. *J. Am. Chem. Soc.*, 82: 376–378.

HOLLAND, H. D., BORCSIK, M., MUNOZ, J. and OXBURGH, U. M., 1963. The coprecipation of Sr²⁺ with aragonite and of Ca²⁺ with strontianite between 90°–100°C. *Geochim. Cosmochim. Acta*, 27: 957–977.

HOLMES, A., 1965. *Principles of Physical Geology*, 2nd ed. Ronald Press, New York, N.Y., 1288 pp.

HOLSER, W. T., KAPLAN, J. R. and SILVERMAN, S. R., 1963. Isotope gechemistry of sulfate rocks. *Ann. Meeting Geol. Soc. Am., New York, Program*, 76: 82 (abstract).

HORIBE, Y. and KOBAYAKAWA, M., 1960. Deuterium abundance of natural waters. *Geochim. Cosmochim. Acta*, 20: 273–283.

HÖRMANN, P., 1962. *Zur Geochemie des Germaniums*. Dissertation, Univ. Tübingen, 44 pp.

HUMMEL, K., 1922. Die Entstehung eisenreicher Gesteine durch Halmyrolyse (= submarine Gesteinszersetzung). *Geol. Rundschau*, 13: 40–81, 97–136.

HUNT, J. M., 1962. *Some Observations on Organic Matter in Sediments*. Paper of the Oil Scientific Session "25 years Hungarian oil, Budapest, October 1962.

HUNT, J. M., STEWART, F. and DICKEY, P. A., 1954. Origin of hydrocarbons of Uinta Basin, Utah. *Bull. Am. Assoc. Petrol. Geologists*, 38: 1671–1698.

ISHIZUKA, T., 1965. *A Paleosalinity Study of the Pliocene Setana Formation of Kuwmatsuna Area, Southern Hokkaido*. Thesis, Hanazawa Univ., (in Japanese).

JACOB, H., 1961. Über bituminöse Schiefer, humose Tone, Brandschiefer und ähnliche Gesteine. *Erdöl Kohle*, 14: 2–11.

JANDA, J. und SCHROLL, E., 1959. Über Borgehalte in einigen ostalpinen Kohlen und anderen Biolithen. *Mineral. Petrog. Mitt.*, 7 (1/2): 118–129.

JOHNS, W. D., 1963. Die Verteilung von Chlor in rezenten marinen und nichtmarinen Sedimenten. *Fortschr. Geol. Rheinland Westfalen*, 10: 215–230.

KAZAKOV, A. V., 1939. Fosfatyne fatsii. (Phosphate facies.) *Trudy NIUIF*, Leningrad–Moskva, 145.

KEAR, D. and ROSS, J. B., 1961. Boron in New Zealand coal ashes. *New Zealand J. Sci.*, 4(2): 360–380.

KEITH, M. L. and DEGENS, E. T., 1959. Geochemical indicators of marine and freshwater sediments. In: P. H. ABELSON (Editor), *Researches in Geochemistry*. Wiley, New York, N.Y., pp. 38–61.

KEITH, M. L., ANDERSON, G. M. and EICHLER, R., 1964. Carbon and oxygen isotopic composition of mollusk shells from marine and fresh water environments. *Geochim. Cosmochim. Acta*, 28: 1757–1786.

KINSMAN, D. J. J., 1965. Coprecipitation of Sr²⁺ with aragonite from sea-water at 15–95°C. *Ann. Meeting. Kansas City, Miss., Geol. Soc. Am., Progr. 1965*, p.87.

KLENOVA, M. V., 1948. *Geologiia Moria*. (Sea geology.) Uchpedgiz, Moscow.

KNEUPER, G., 1964. Grundzüge der Sedimentation und Tektonik im Oberkarbon des Saarbrückes Hauptsattels. *Oberrhein. Geol. Abh.*, 13: 1–49.

KOCZY, F. F., TOMIC, E. und HECHT, F., 1957. Zur Geochemie des Urans im Ostseebecken. *Geochim. Cosmochim. Acta*, 11: 85–102.

KOCZY, F. F., ANTAL, P. S. und JOENSUU, O., 1953. Die natürlichen radioaktiven

Elemente in Sedimenten. *Fortschr. Geol. Rheinland Westfalen*, 10: 201–214.
KREJCI-GRAF, K., 1962a. Über Ölfeldwässer. *Erdöl Kohle*, 15: 102–109.
KREJCI-GRAF, K., 1962b. Über Bituminierung und Erdölentstehung. *Freiberger Forschungsh.*, *C*, 123: 5–34.
KREJCI-GRAF, K., 1963a. Organische Geochemie. *Naturw. Rundschau*, 16: 175–186.
KREJCI-GRAF, K., 1963b. Unterscheidungsmöglichkeiten mariner und nichtmariner Sedimente, D. Rückblick. Stand der gegenwärtigen Untersuchungen und offene Fragen. *Fortschr. Geol. Rheinland Westfalen*, 10: 449–462.
KREJCI-GRAF, K., 1963c. Diagnostik der Salinitätsfazies der Ölwässer. *Fortschr. Geol. Rheinland Westfalen*, 10: 367–448.
KREJCI-GRAF, K., 1966. Geochemische Faziesdiagnostik. *Freiberger Forschungsh.*, *C*, 224: 80 pp.
KREJCI-GRAF, K. und LEIPERT, TH., 1936. Bromgehalte in mineralischen, kohligen und bituminösen Ablagerungen. *Z. Prakt. Geol.*, 44: 117–123.
KREJCI-GRAF, K. und ROMEIS, H., 1962. Ionen-Abgabe aus Peliten. *Chem. Erde*, 22: 371–385.
KREJCI-GRAF, K. und WICKMAN, F. E., 1960. Ein geochemisches Profil durch den Lias alpha. *Geochim. Cosmochim. Acta*, 18: 259–272.
KREJCI-GRAF, K., KLEIN, K., KREHER, A., ROSSWURM., M. und WENZEL, G., 1965. Versuche zur geochemischen Fazies-Diagnostik. *Chem. Erde*, 24: 115–146.
KREJCI-GRAF, K., APPELT, W. und KREHER, A., 1966. Zur Geochemie des Wiener Beckens. *Geol. Mitt.*, 7: 49–108.
KROMER, H., 1963. *Untersuchungen über den Mineralbestand des Knollenmergel-Keupers in Württemberg*. Dissertation, Univ. Tübingen, 71 pp.
KRUMBEIN, W. C. and SLOSS, L. L., 1951. *Stratigraphy and Sedimentation*. Freeman, San Francisco, Calif., 497 pp.
KRUMM, H., 1963. Der Tonmineralbestand fränkischer Keuper- und Juratone unterschiedlicher Entstehung. *Fortschr. Geol. Rheinland Westfalen*, 10: 327–330.
KÜBLER, B., 1963. Untersuchungen über die Tonfraktion der Trias der Sahara: Ein Beispiel gegenseitiger Abhängigkeit der Salinität und der Tonminerale. *Fortschr. Geol. Rheinland Westfalen*, 10: 319–324.
KULP, J. L., 1951. Origin of the hydrosphere. *Bull. Geol. Soc. Am.*, 62: 326–329.
KULP, J. L., TUREKIAN, K. and BOYD, D. W., 1952. Strontium content of limestones and fossils. *Bull. Geol. Soc. Am.*, 63: 701–716.
KUMMEL, B., 1957. Paleocology of Lower Triassic formations of southeastern Idaho and adjacent areas. *Geol. Soc. Am.*, *Mem.*, 67 (2): 437–467.
KURODA, P. K. and SANDELL, E. B., 1950. Determination of chlorine in silicate rocks. *Anal. Chem.*, 22; 1144–1145.

LANDERGREN, S., 1945. Contribution to the geochemistry of boron. *Ark. Kem. Mineral. Geol.*, 19(5A)-25: 7 pp.; 26: 31 pp.
LANDERGREN, S., 1958. On the distribution of boron on different size classes in marine clay sediments. *Geol. Fören. Stockholm Förh.*, 80 (492): 14–107.
LANDERGREN, S. und MANHEIM, F. T., 1963. Über die Abhängigkeit der Vertei-

lung von Schwermetallen von der Fazies. *Fortschr. Geol. Rheinland Westfalen*, 10: 173–192.

LEUTWEIN, F., 1963. Spurenelemente in rezenten Cardien verschiedener Fundorte. *Fortschr. Geol. Rheinland Westfalen*, 10: 283–292.

LEUTWEIN, F. und RÖSLER, H. J., 1956. Geochemische Untersuchungen and paläozoischen und mesozoischen Kohlen Mittel- und Ostdeutschlands. *Freiberger Forschungsh.*, C, 19: 196 pp.

LEVINSON, A. A. and LUDWICK, J. C., 1966. Speculation on the incorporation of boron into argillaceous sediments. *Geochim. Cosmochim. Acta*, 30: 855–861.

LIVINGTON, D. A., 1963. The sodium cycle and the age of the ocean. *Geochim. Cosmochim. Acta*, 27: 1055–1069.

LOCHMAN, C., 1956. Stratigraphy, paleontology, and paleogeography of the *Elliptocephala* asaphoides strata in the Cambridge and Hoosick quadrangles, New York. *Bull. Geol. Soc. Am.*, 67: 1331–1396.

LOTZE, F., 1957. *Steinsalz und Kalisalze*, 1. Borntraeger, Berlin, 465 pp.

MÄDLER, K., 1963. Charaphyten und Halophyten. *Fortschr. Geol. Rheinland Westfalen*, 10: 121–128.

MASON, B., 1952. *Principles of Geochemistry*. Wiley, New York, N.Y., 276 pp.

MAUCHER, A., 1962. *Die Lagerstätten des Urans*. Vieweg, Braunschweig, 162 pp.

MICHALIČEK, M., 1961. Geochemie der Tiefwässer und Ablagerungen des westlichen Teils der Kleinen Donau-Tiefebene. *Konf. Nauk. Inst. Naft. Polski, 2, Czechoslowacji i Wegier*, pp.171–181.

MILLOT, G., 1949. Relations entre la constitution et la genèse des roches sédimentaires argileuses. *Géol. Appl. Prosp. Min.*, 2: 1–352.

MILLOT, G., 1964. *Géologie des Argiles*. Masson, Paris, 499 pp.

MOORE, R. C., 1948. Stratigraphical paleontology. *Bull. Geol. Soc. Am.*, 59: 301–325.

MOORE, R. C., 1949. Meaning of facies. *Geol. Soc. Am., Mem.*, 39: 1–34.

MOORE, R. C., 1957. Modern methods of paleoecology. *Bull. Am. Assoc. Petrol. Geologists*, 41: 1775–1801.

MOTOJIMA, K., ANDO, A. and KAWANO, M., 1960. A study of sedimentary rock. Chemical composition and environment of sedimentation. *J. Japan. Assoc. Petrol. Tech.*, 25: 298–303.

MÜLLER, G., 1964. *Methoden der Sedimentuntersuchung*. Schweizerbart, Stuttgart, 303 pp.

MÜLLER, G., 1967. Diagenesis in argillaceous sediments. In: G. LARSEN and G. V. CHILINGAR (Editors), *Diagenesis in Sediments*. Elsevier, Amsterdam, pp.127–177.

MÜLLER, G., 1969. Sedimentary phosphate method for estimating paleosalinities: limited applicability. *Science*, 163: 812–813.

MÜLLER, G. and FÖRSTNER, U., 1968. General relationship between suspended sediment concentration and water discharge in the Alpenrhein and some other rivers. *Nature*, 217 (5125): 244–245.

MÜLLER, G. und HAHN, C., 1964. Schwermineral- und Karbonatführung der Fluss-Sande im Einzugsgebiet des Alpenrheins. *Neues Jahrb. Mineral., Monatsh.*, 1964: 371–375.

MÜLLER, G., NIELSEN, H. und RICKE, W., 1966. Schwefelisotopen-Verhältnisse in Formationswässern und Evaporiten Nord- und Süddeutschlands. *Chem. Geol.*, 1: 211–220.

NAKAI, N. and JENSEN, M. L., 1960. Biochemistry of sulfur isotopes. *J. Earth Sci., Nagoya Univ.*, 8: 181–196.

NICHOLLS, G. D., 1963. Environmental studies in sedimentary geochemistry. *Sci. Progr. London*, 51: 12–31.

NIELSEN, H., 1965. Schwefelisotope im marinen Kreislauf und das δ ^{34}S der früheren Meere. *Geol. Rundschau*, 55: 160–172.

NIELSEN, H. und RICKE, W., 1964. Schwefelisotopen-Verhältnisse von Evaporiten aus Deutschland; ein Beitrag zur Kenntnis von δ ^{34}S im Meerwasser-Sulfat. *Geochim. Cosmochim. Acta*, 28: 577–591.

NOAKES, J. E. and HOOD, D. W., 1961. Boron–boric acid complexes in seawater. *Deep-Sea Res.*, 8: 121–129.

NELSON, B. W., 1967. Sedimentary phosphate method for estimating paleosalinities. *Science*, 158: 917–920.

ODUM, H. T., 1950. *Biogeochemistry of Strontium*. Thesis, Yale Univ.

ODUM, H. T., 1957. Biochemical deposition of strontium. *Texas Univ., Inst. Marine Sci.*, 4: 39–144.

OPARIN, A. J., 1957. *The Origin of Life on the Earth*, 3rd. ed. Acad. Press, New York, N.Y., 495 pp.

OSTLUND, G., 1959. Isotopic composition of sulfur in precipitation and seawater. *Tellus*, 11: 478–480.

OTTE, M. U., 1953. Spurenelemente in einigen deutschen Steinkohlen. *Chem. Erde*, 16: 239.

PACHAM, G. H. and CROOK, K. A. W., 1960. The principle of diagenetic facies and some of its implications. *J. Geol.*, 68: 392–407.

PAPP, A., 1963. Das Verhalten neogener Molluskenfaunen bei verschiedenen Salzgehalten. *Fortschr. Geol. Rheinland Westfalen*, 10: 35–48.

PAVLENKO, I. A. and GAVRILOVA, I. P., 1964. Heavy metals content of loose sediments in the upper part of the Tanalyk River Basin, southern Urals. *Intern. Geol. Rev.*, 6: 18–34.

PILKEY, O. H. and GOODELL, H. G., 1964. Comparison of the composition of fossil and recent mollusk shells. *Geol. Soc. Am. Bull.*, 75: 217–228.

PINSAK, A. P. and MURRAY, H. H., 1960. Regional clay-mineral patterns in the Gulf of Mexico. *Proc. Natl. Conf. Clays Clay Minerals, 7th—Natl. Acad. Sci. Natl. Res. Council, Publ.*, pp.162–177.

PLASS, G. N., 1956. Effect of carbon dioxide variations on climate. *Am. J. Phys.*, 24: 376–387.

PLUMSTEAD, E. P., 1961. The Permo-Carboniferous coal measures of the Transvaal South Africa—an example of the contrasting stratigraphy in the southern and northern hemispheres. *Congr. Avan. Études Stratigraph. Géol. Carbonifère, Compt. Rend.*, 4, *Heerlen, 1958*, 2: 545–550.

PORRENGA, D. H., 1963. Bor in Sedimenten als Indiz für den Salinitätsgrad. *Fortschr. Geol. Rheinland Westfalen*, 10: 267–270.

POTONIE, R., 1928. *Allgemeine Petrographie der "Ölschiefer" und ihrer Verwand-*

ten, mit Ausblick auf die Erdölentstehung. Bornträger, Berlin, 173 pp.

POWERS, M. C., 1954. Clay diagenesis in the Chesapeake Bay area. *Proc. Natl. Conf. Clays Clay Minerals, 2nd—Natl. Acad. Sci. Natl. Res. Council, Publ.,* pp.68–80.

PRASHNOWSKY, A. A., 1963. Verteilung von organische Substanzen in Sedimenten. *Fortschr. Geol. Rheinland Westfalen,* 10: 295–306.

PROKOFIEV, W. A., 1964. Elementare chemische Zusammensetzung der Schalen von palaeozoischen Brachiopoden. *Geochimija,* 1: 75–81.

PUSTOVALOV, L. V., 1940. *Petrografiia Osadochnykh Porod.* (Petrography of sedimentary rocks). Gostoptekhizdat, Leningrad–Moskva.

RANKAMA, K., 1954. *Isotope Geology.* Pergamon, London, 535 pp.

RANKAMA, K., 1963. *Progress in Isotope Geology.* Wiley, London-New York, 705 pp.

RANKAMA, K. and SAHAMA, TH. G., 1950. *Geochemistry.* Chicago Univ. Press, Chicago, Ill., 912 pp.

REINKENSMEIER, U., 1970. *Restgase in Kohle und Nebengestein des Saarkarbons.* Dissertation, Univ. Tübingen, im Druck

REMANE, A., 1963. Biologische Kriterien zur Unterscheidung von Süss- und Salzwassersedimenten. *Fortschr. Geol. Rheinland Westfalen,* 10: 9–34.

REYNOLDS, R. C., 1965a. The concentration of boron in Precambrian seas. *Geochim. Cosmochim. Acta,* 29: 1–16.

REYNOLDS, R. C., 1965b. Boron and oceanic evolution: a reply. *Geochim. Cosmochim. Acta,* 29: 1008–1009.

RICHARDS, F. A. and BENSON, B. B., 1961. Nitrogen/argon nitrogen isotope ratios. *Deep-Sea Res.,* 7: 254–264.

RICHTER, G., 1941. Geologische Gesetzmässigkeiten in der Metallführung des Kupferschiefers. *Arch. Lagerstätt. Forschung,* 73: 61 pp.

RICHTER-BERNBURG, G., 1955. Über salinare Sedimentation. *Z. Deut. Geol. Ges.,* 105: 593–645.

RICKE, W., 1963. Geochemie des Schwefels und ihre Anwendung auf Faziesprobleme. *Fortschr. Geol. Rheinland Westfalen,* 10: 271–278.

RITTENBERG, S. C., EMERY, K. O. HÜLSEMANN, J., DEGENS, E. T., FAY, R. C., REUTER, J. H., GRADY, J. R., RICHARDSON, S. H. and BRAY, E. E., 1963. Biochemistry of sediments in experimental mohole. *J. Sediment. Petrol.,* 33: 140–172.

RONOV, A. B., 1959. On the post-Cambrian geochemical history of the atmosphere and hydrosphere. *Geochimija,* 5: 493–506 (English transl.).

RONOV, A. B. and MIGDISOV, A. A., 1960. On the relationship between normal (Clarke) and ore concentrations of alumina in the sedimentary cycle. *Intern. Geol. Congr., 21st., Copenhagen, 1960, Geochem. Cycle,* 1: 157–177.

ROSENFELD, W. D. and SILVERMAN, S. R., 1959. Carbon isotope fractionation in bacterial production of methane. *Science,* 130: 1658–1659.

ROTH, J., 1879–90. *Allgemeine und chemische Geologie,* 1–3. Hertz, Berlin, 633/695/530 pp.

ROTTGARDT, D., 1952. Mikropaläontologisch wichtige Bestandteile rezenter brackischer Sedimente an den Küsten Schleswig-Holsteins. *Meyniana,* 1: 169–228.

RUBEY, W. W., 1951. Geologic history of sea water. *Bull. Geol. Soc. Am.*, 62: 1111–1147.

RUBEY, W. W., 1955. Development of the hydrosphere and atmosphere with special reference to probable composition of the early atmosphere. In: *Crust of the Earth—Geol. Soc. Am., Spec. Papers*, 62: 631–650.

RUKHIN, L. B., 1948. *Tipy Peschahykh Fatsii*. (Types of sandy facies.) Litolog. Gostoptekhizdat, Leningrad–Moskva.

SAHAMA, TH. G., 1945. Spurenelemente der Gesteine im südlichen Finnisch Lappland. *Bull. Comm. Geol. Finlande*, 135: 86 pp.

SCHROLL, E., 1961. Seltene Elemente in biogenen Sedimenten. *Mineral. Petrog. Mitt.*, 7: 488–490.

SCHULZE, G., 1960. Stratigraphische und genetische Deutung der Bromverteilung in den mitteldeutschen Steinsalzlagern des Zechsteins. *Freiberger Forschungsh.*, C83: 114 pp.

SCHUSTER, A., 1963. Konnektierung von Bohrlochmessungen im Westfal der Bohrung Münsterland 1 mit Messungen anderer Bohrungen. *Fortschr. Geol. Rheinland Westfalen*, 11: 487–516.

SCHUSTER, A., 1968. Karbonstratigraphie nach Bohrlochmessungen. *Erdöl Erdgas Z.*, 84: 439–457.

SCHWARZBACH, M., 1961. *Das Klima der Vorzeit. Eine Einführung in die Paläoklimatologie*. Enke, Stuttgart, 275 pp.

SEIBOLD, E., MÜLLER, G. und FESSER, H., 1958. Chemische Untersuchungen eines Sapropels aus der mittler Adria. *Erdöl Kohle*, 11: 296–300.

SEIDEL, G., 1962. Zusammenhänge zwischen Rotfazies und orogenen Bodenbewegungen im Südosten der subvariszischen Saumtiefe. *Congr. Avan. Études Stratigraph. Géol. Carbonifère, Compt. Rend.*, 4, Heerlen, 1958, 3: 609–619.

SHERBAKOV, A. V., 1961. *Gidroglokhi-micheskie Issledovaniya pri Poiskakh i Razvedke Podzemuykh Boronosnykh Vodi*. Vsesoyuz. Nauch, Issled. Inst. Gidrogeol. i. Inzh. Geol., Moskva, 127 pp.

SIEVER, R., BECK, K. C. and BERNER, R. A., 1965. Composition of interstitial waters of modern sediments. *J. Geol.*, 73: 39–1965.

SILVERMAN, S. R., 1951. The isotope geology of oxygen. *Geochim. Cosmochim. Acta*, 2: 26–42.

SILVERMAN, S. R. and EPSTEIN, S., 1958. Carbon isotope compositions of petroleum and other sedimentary organic materials. *Bull. Am. Assoc. Petrol. Geologists*, 42: 998–1012.

SINGH, J. B., 1966. Borgehalsbestimmungen im Knollenmergel-Keuper (Südwest Deutschland). *Chem. Geol.*, 1: 251–258.

SLOSS, L. L., KRUMBEIN, W. C. and DAPPLES, E. C., 1949. Integrated facies analysis. *Geol. Soc. Am., Mem.*, 39: 91–123.

SKOPINTSEV, B. A. and TIMOFEYEVA, S. N., 1960. The content of organic carbon in the water of the northeastern part of the Atlantic Ocean. *Dokl. Earth Sci. Sect.*, 133: 641–643 (transl. from Russian).

SPEARS, D. A., 1965. Boron in some British Carboniferous sedimentary rocks. *Geochim. Cosmochim. Acta*, 29: 315–328.

SPJELDNAES, N., 1962. Boron is some Norwegian paleozoic sediments. *Norsk Geol. Tidsskr.*, 42(1/2): 191–195.

138 REFERENCES

STADLER, G., 1963. Zusammenhänge zwischen Mineralfazies und Borgehalten. *Fortschr. Geol. Rheinland Westfalen*, 10: 325–326.

STADNIKOFF, G., 1958. Ein chemisches Verfahren zur Feststellung der Ablagerungsbedingungen von Tonen und tonigen Gesteinen. *Glückauf*, 94: 58–62.

STAVROV, O. D. and KHITROV, V. G., 1962. Possible geochemical relationship observed between cesium and boron. *Geochemistry (U.S.S.R.)*, 1962, pp.57–67 (transl. from Russian).

STRØM, K., 1948. A concentration of uranium in black muds. *Nature*, 162: 162.

SWAN, E. F., 1956. The meaning of strontium/calcium ratios. *Deep-Sea Res.*, 4: 71.

SWAIN, F. M., 1961. Limnology and amino-acid content of some lake deposits in Minnesota, Montana, Nevada and Louisiana. *Geol. Soc. Am. Bull.*, 72: 519–546.

SWAINE, D. J., 1962. Boron in New South Wales Permian coals. *Australian J. Sci.*, 25(6): 265–266.

SWAINE, F. M., BLUMENTALS, A. and PROKOPOVICH, N., 1958. Bituminous and other organic substances in Precambrian of Minnesota. *Bull. Am. Assoc. Petrol. Geologists*, 42: 173–189.

TEIS, R. V., SCHUPAKIN, N. S. und NAIDIN, D. P., 1957. Paläotemperaturbestimmung nach der Isotopenzusammensetzung der Sauerstoffs im Calcit der Schalen einiger Versteinerungen der Krim. *Geochimija*, 4: 271 (in Russian).

THODE, H. G., 1949. Variation in abundances of isotopes in nature. *Research (London)*, 2: 154–161.

THODE, H. G. und MONSTER, J., 1964. S-Isotopenverhältnisse in Evaporiten und in den früheren Ozeanen. In: V. A. VINOGRADOV (Redakteur-Herausgeber), *Khimiya zemnoskory, Trudy Geokhimicheskoi Konferentsii Posvyashchennoi Stoletiyn so Dnya Rozhdeniya Akadimika.—J. Zd. Akad. Nauk S.S.S.R.*, Moskau, 2: 589–600 (Russ.).

THODE, H. G., MONSTER, J. and DUNFORD, H. B., 1961. Sulphur isotope geochemistry. *Geochim. Cosmochim. Acta*, 25: 159–174.

THOMPSON, T. G. and CHOW, T. J., 1956. The strontium/calcium ratio in carbonate-secreting marine organisms. *Papers Marine Biol. Oceanog.*, *Suppl.*, 3: 20–39.

TOURTELOT, H. A., 1964. Minor-element composition and organic carbon content of marine and non-marine shales of Late Cretaceous age in the western interior of the United States. *Geochim. Cosmochim. Acta*, 28:1579–1604.

TOURTELOT, H. A., SCHULTZ, L. G. and HUFFMAN, C., 1961. Boron in bentonite and shale from the Pierre Shale, South Dakota, Wyoming and Montana. *U.S., Geol. Surv., Proffess. Papers*, 424C: 288–292.

TRASK, P. D., 1932. *Origin and Environment of Source Sediments of Petroleum.* Am. Petroleum Inst., Houston, Texas, 323 pp.

TREIBS, A., 1934. Chlorophyll- und Häminderivate in bituminösen Gesteine, Erdölen, Erdwachsen und Asphalten. Ein Beitrage zur Entstehung des Erdöls. *Ann. Chem.*, 510: 42–62.

TREIBS, A., 1935. Chlorophyll- und Häminderivate in bituminösen Gesteine, Erdölen, Kohlen, Phosphoriten. *Ann. Chem.*, 517: 172.

TRUSHEIM, F., 1957. Über Halokinese und ihre Bedeutung für die strukturelle Entwicklung Norddeutschlands. *Z. Deut. Geol. Ges.*, 109: 111–151.

TUREKIAN, K. K., 1959. The terrestrial economy of helium and argon. *Geochim. Cosmochim. Acta*, 17: 37–43.

TUREKIAN, K. K., 1964. The marine geochemistry of strontium. *Geochim. Cosmochim. Acta*, 28: 1479–1496.

TUREKIAN, K. K. and KULP, J. L., 1956. The geochemistry of strontium. *Geochim. Cosmochim. Acta*, 10: 245–296.

UREY, H. C., 1947. The thermodynamic properties of isotopic substances. *J. Chem. Soc.*, 562–581.

UREY, H. C., 1951a. The origin and development of the earth and other terrestrial planets. *Geochim. Cosmochim. Acta*, 1: 209–277.

UREY, H. C., 1951b. Measurement of paleotemperatures. *Bull. Geol. Soc. Am.*, 62: 399–416.

UREY, H. C. 1959. *Handbuch der Physik*, 52. Springer, Berlin–Heidelberg–Göttingen, 601 pp.

UREY, H. C., EPSTEIN, S., MCKINNEY, C. R. and MCCREA, J., 1948. Method of measurement of paleotemperatures. *Bull. Geol. Soc. Am.*, 59: 1359–1360 (abstract).

UREY, H. C., LOWENSTAM, S., EPSTEIN, S. and MCKINNEY, C. R., 1951. Measurement of paleotemperature and temperatures of the Upper Cretaceous of England, Denmark, and the southeastern United States. *Bull. Geol. Soc. Am.*, 62: 399–416.

VALLENTYNE, J. R., 1957. *Annual Report of the Director of the Geophysical Laboratory, 1956–1957*. Carnegie Inst. Washington, D.C., 56: 185.

VASSOYEVICH, N. B., 1948. *Evolutsiia Predstavlenii o Geologischeskikh Fatsiiahk.* (Evolution of concepts of geologic facies.) Gostoptekhizdat, Leningrad–Moskva.

VINOGRADOV, A. P., 1953. The elementary chemical composition of marine organisms. *Sears Found. Marine Res.*, *Yale Univ.*, *Mem.*, 2: 647 pp.

VINOGRADOV, A. P., 1957. Variation in the chemical composition of carbonate rocks of the Russian platform. *Geochim. Cosmochim. Acta*, 12: 273–276.

VINOGRADOV, A. P. and RONOV, A. B., 1956. Evolution of the chemical composition of the Russian Platform clays. *Geochimija*, 2: 3–18.

VINOGRADOV, A. P. and TEIS, R. V., 1941. Isotopic composition of oxygen of different origin. *Compt. Rend. Acad. Sci. S.S.S.R.*, 33: 490 (in Russian).

VON ENGELHARDT, W., 1960. *Mineralogie und Petrographie in Einzeldarstellungen, 2. Der Porenraum der Sedimente.* Springer, Berlin–Heidelberg–Göttingen, 207 pp.

VON ENGELHARDT, W., 1967. Interstitial solutions and diagenesis. In: G. LARSEN and G. V. CHILINGAR (Editors), *Diagenesis in Sediments.* Elsevier, Amsterdam. pp. 503–521.

VON ENGELHARDT, W., FÜCHTBAUER, H. und LEMCKE, K., 1953. Geologische und sedimentpetrographische Untersuchungen im Westteil der ungefalteten Molasse des süddeutschen Alpenvorlandes. *Geol. Jahrb., Beih.*, 11: 190 pp.

VON ENGELHARDT, W., MÜLLER, G. und KROMER, H., 1962. Dioktaedrischer

140   REFERENCES

Chlorit (Sudoit) in Sedimenten des Mittleren Keupers von Plochingen (Württ.). *Naturwissenschaften*, 49: 205–206.

VON GAERTNER, H. R. und KROEPELIN, H., 1956. Petrographische und chemische Untersuchungen am Posidonienschiefer Nordwestdeutschlands. *Erdöl Kohle*, 9/10: 588–592, 680–682.

WALKER, C. T., 1964. Paleosalinity in Upper Visean Yoredale formation of England (geochemical method for locating porosity). *Bull. Am. Assoc. Petrol. Geologists*, 48: 207–220.

WALKER, C. T. and PRICE, N. B., 1963. Departure curves for computing palaeosalinity from boron in illites and shales. *Bull. Am. Assoc. Petrol. Geologists*, 47: 833–841.

WALTHER, J., 1893. *Einleitung in die Geologie als historische Wissenschaft. 1. Bionomie. 2. Die Lebensweise der Meerestiere*. Fischer, Jena, 531 pp.

WASMUND, E., 1930. Bitumen, Sapropel und Gyttja. *Geol. Fören. Stockholm Förh.*, 52: 315–350.

WASMUND, E., 1935. Die Bildung von anabituminösem Leichenwachs unter Wasser. *Schriften Brennstoff Geol.*, 10: 1–70.

WEAVER, C. E., 1958. Geologic interpretation of argillaceous sediments, 1. Origin and significance of clay minerals in sedimentary rocks. *Bull. Am. Assoc. Petrol. Geologists*, 42: 254–271.

WEBER, J. N., 1964a. Paleoenenvironmental significance of carbon isotopic composition of siderite nodules. *J. Sediment. Petrol.*, 34: 814–818.

WEBER, J. N., 1964b. Palaeoclimatic significance of δ-oxygen-18 time trends observed by oxygen isotopic analysis of freshwater limestones. *Nature*, 203: 969–970.

WEBER, V., 1958. Facies of deposits favourable for the formation of bitumen. *Eclogae Geol. Helv.*, 51: 617–622.

WEDEPOHL, K. H., 1963. Einige Überlegungen zur Geschichte des Meerwassers. *Fortschr. Geol. Rheinland Westfalen*, 10: 129–150.

WEDEPOHL, K. H., 1964. Untersuchung an Kupferschiefer in Nordwestdeutschland. *Geochim. Cosmochim. Acta*, 28: 305–364.

WEEKS, L. G. (Editor), 1958. *Habit of Oil*. Am.Assoc. Petrol. Geologists, Tulsa, Okla., 1384 pp.

WELLS, J. W., 1944. Middle Devonian bone beds of Ohia. *Bull. Geol. Soc. Am.*, 55: 273–302.

WERNER, H., 1963. Über das Calcium/Magnesium-Verhältnis in Torf und Kohle. *Fortschr. Geol. Rheinland Westfalen*, 10: 279–282.

WELTE, D. H., 1959. Geochemische Untersuchungen von organischen Substanzen aus oberkarbonischen Tonschiefern mariner und limnischer Fazies. *Neues Jahrb. Geol. Paläontol., Monatsh.*, 1959, p.84.

WHITEHOUSE, U. G. and McCARTER, R. S., 1958. Diagenetic modification of clay-mineral types in artificial sea water. *Proc. Natl. Conf. Clays Clay Minerals, 5th—Natl. Acad. Sci. Natl. Res. Council, Publ.*, pp.84–119.

WHITEHOUSE, U. G., JEFFREY, L. M. and DEBBRECHT, J. D., 1960. Differential settling tendencies of clay minerals in saline waters. *Proc. Natl. Conf. Clays Clay Minerals, 7th—Natl. Acad. Sci. Natl. Res. Council, Publ.*, pp.1–79.

WICKMANN, F. E., 1953. Wird das Häufigkeitsverhältnis der Kohlenstoffisotopen bei der Inkohlung verändert? *Geochim. Cosmochim. Acta*, 3: 244–252.

WINTER, J., 1969. Vulkanogene Bentonitlagen im Oberemsium und Eifelium der Eifel. *Z. Deut. Geol. Ges.*, im Druck.

Index

Date Due